A Focus on Peatlands and Peat Mosses

 GREAT LAKES ENVIRONMENT

Warren H. Wagner, Jr., Series Editor

A Focus on Peatlands and Peat Mosses
    Howard Crum

A Focus on
Peatlands and Peat Mosses

Howard Crum

In Collaboration with Sandra Planisek

Ann Arbor
THE UNIVERSITY OF MICHIGAN PRESS

First paperback edition 1992
Copyright © the University of Michigan 1988
All rights reserved
Published in the United States of America by
The University of Michigan Press
Manufactured in the United States of America

1995 1994 1993 1992   5 4 3 2

Library of Congress Cataloging-in-Publication Data

Crum, Howard Alvin. 1922–
    A focus on peatlands and peat mosses / Howard Crum in collaboration with Sandra Planisek.
        p.   cm.—(Great Lakes environment)
    Bibliography: p.
    Includes index.
    ISBN 0-472-09378-9 (cloth : alk.) — ISBN 0-472-06378-2 (paper : alk.)
    1. Peatland plants—Ecology. 2. Peatland plants—Middle West—Ecology. 3. Peatlands. 4. Peatlands—Middle West. 5. Peat mosses. 6. Peat mosses—Middle West   I. Planisek, Sandra, 1948–.
II. Title.   III. Series.
QK938.P42C78   1988
574.5'26325—dc19
                                                                   88-4768
                                                                     CIP

Preface

> No man ever followed his genius until it misled him.
> —Thoreau

We have gone to peatlands in search of plants and especially mosses. We have walked with delight on quaking mats, and we have fallen through them. We have rejoiced in the beauty of bogs and marveled at the way plants make a living there in the face of grudging hospitality. We have wondered why every fen and every bog is somehow unique and yet follows similar patterns of succession. Rich fens, high in calcium, give rise to white cedar swamps where peat accumulation is limited. But fens of lesser richness build up enough peat to alter the movement and chemistry of water, eventually to the extent that they are transformed into acid bogs receiving water and nutrients only from the atmosphere. In the change from mineral-rich to mineral-poor peatlands, *Sphagnum* leads the way, from sedgy intermediate fens through wet *Sphagnum* lawns, or poor fens, to open bogs covered by a low-shrub, hummock-hollow complex and eventually to black-spruce muskeg.

We have been especially interested in defining the species of *Sphagnum* in terms of the meaningful roles they play in the ecology of peatlands. We intended to put together a small book to show how the peat mosses contribute to their environment. We hoped to show how the species can be recognized in the field and used as environmental indicators. But as we grew in knowledge, the book became a peatland primer, a student's guide to the habitats where *Sphagnum* abounds. It is not a scholarly treatise. It is the gist of what we know, what we have seen, heard, and read. It is by no means exhaustive. Nor are peatland studies as advanced as one might wish. The population ecology of peatlands, for example, needs attention. We have little or no information on reproductive biology, dispersal, seed germination, establishment of seedlings, and population genetics. Nor have we put together the available information on the animals that live in peatland habitats.

The literature on peatlands is endlessly informative. But the detailed information on the vegetation, hydrology, and chemistry at one place scarcely translates into generalizations useful in another. Confusion is compounded by the language of Babel. Even specialists in peatland ecology fail to make meaningful distinctions between, let us say, marshes and fens or poor fens and bogs. The word bog can mean a rich, marly fen or an entire fen-to-bog successional series. The din of language obscures the theme of orderly progression of vegetational changes leading up to a climax. But is there a climax? Is succession orderly and directional, or is it something done and undone over and over again?

This book has a definite Midwest bias, but in our attempt to put the peatlands of the Upper Midwest into focus, we have tried to give a glimpse at peatlands elsewhere. We have explored the ways in which peatlands are and should be classified, as well as their changes through time. We have taken notice of the plants associated with those changes and the possibilities of nutrient cycling as fens change to bogs. We have presented a simplified, perhaps simplistic account of peatland archives in which bygone plant responses to climates are recorded. The need for the wise use of peatland resources we have also tried to make clear. We have shared our knowledge of the peat mosses, not merely because of bryological bias, but also because of their basic importance in the fen and bog environments. Chapter 5 is intended, in part, to show the importance of peat mosses in the ever-changing vegetation of peatlands and also to give some sense of species definitions, by "look" and by habitat niche. The final chapter is a fairly technical exposition on the taxonomy of peat mosses, useful to the uninitiated field biologist, if only because of the illustrations, and also to the bryological user.

Our bibliography gives reference to many sources of information other than those specifically cited. We draw attention here to several general accounts of peatlands especially contributory to basic understandings and, for the most part, especially readable:

Andrus (1986)	Gates (1942b)
Boelter and Verry (1977)	Godwin (1978, 1981)
Conway (1949)	Heinselman (1963, 1965, 1970)

Johnson (1985)
Larsen (1982)
Lucas (1982)
Moore and Bellamy (1973)
Pollett, Rayment, and Robertson (1979)

Rigg (1940–51)
Transeau (1905–6)
Worley (1981)

We can scarcely give thanks to all the persons who contributed to our joy of learning. Norton G. Miller, Lewis E. Anderson, Wolfgang Maass, Dale H. Vitt, Harold Robinson, Jerry Snider, Barbara Madsen, Richard Futyma, and Guy Brassard were stout companions in the field. Helpful criticisms were provided by Edward Voss, Anton Reznicek, Terry Sharik, Cyrus McQueen, Michael Penskar, Dennis Albert, James Larsen, Ian Worley, Wilfred Schofield, and Norton Miller, as well as William Buck, Jerry Snider, Robert Naczi, and Sarah Hoot. We got good photographs from Doyle Wells of the Canadian Forestry Service, Charles Tarnocai of the Canadian Department of Agriculture, Lynda Dredge of the Canadian Geological Survey, and Harry Tyler of the Critical Areas Program of the State of Maine. Jeffrey Holcombe and David Levick provided many light and scanning electron microscope photographs showing the structure and life history of *Sphagnum*. Robert Ireland did us many favors throughout the process of writing. Peter Kilham was a dependable source of limnological information. The drawings of *Sphagnum*, prepared by Constance Butley, were made available for republication by the New York Botanical Garden. The drawings of higher plants were prepared by Marie Wohadlo and the charts and diagrams by Sarah Hoot.

Hidden subsidies came from a number of NSF grants in support of past work on *Sphagnum* taxonomy. Research time and facilities were provided by the National Museum of Canada and the Herbarium, the Department of Biology, and the Biological Station at the University of Michigan.

Contents

Chapter
1. Fens and Bogs, the Peat-Storing Wetlands   1
2. The Origin and Development of Peatlands   29
3. Peatland Plants and the Tonic of Wildness   57
4. Nutrients in Scant Supply   111
5. Life in a Wet Desert   141
6. Peatland Archives   163
7. The Nature and Use of Peat and Peatland   177
8. A Close Look at Peat Mosses: A Bryologist's Vademecum   201

Glossary   277

Bibliography   285

Index   299

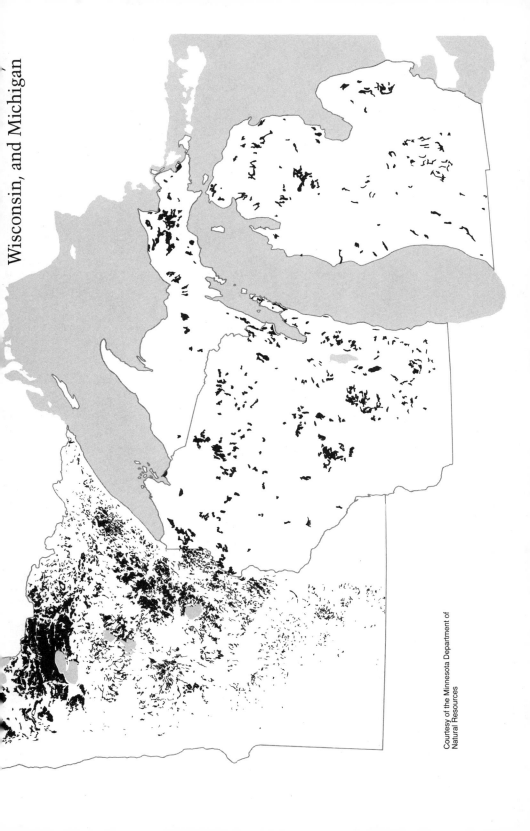

Wisconsin, and Michigan

Courtesy of the Minnesota Department of Natural Resources

Chapter 1

Fens and Bogs, the Peat-Storing Wetlands

> If the day and night are such that you greet them with joy, and life emits a fragrance like flowers and sweet scented herbs, is more elastic, more starry, more immortal—that is your success. All Nature is your congratulation and you have cause momentarily to bless yourself.
> —Thoreau

Boggy wetlands are a joy and a trial. The farmer views a soggy piece of land rather differently from a biological enthusiast. A biologist with a view toward conservation has concerns to convey to the taxpayer, but can the taxpaying farmer understand a biologist who calls swamps bogs and bogs swamps? Those and other terms related to wetlands need, yet defy, definition.

Peatlands develop in places water-soaked throughout the growing season. Bogs and fens are peatlands, but not all wetlands, boggy underfoot, are peat storing. It is only when plant growth exceeds decomposition that organic matter accumulates as sodden peat. *Fens* (fig. 1-1) develop under the influence of mineral-rich ground or surface water. They are characterized by grassy plants, mostly sedges. They can be succeeded by coniferous swamps or by bogs, depending on water movement and water quality. Conditions favorable to the continued growth of *Sphagnum* lead toward a bog sequence. *Bogs* (figs. 1-2–3) are mineral-poor, acid peatlands raised above the groundwater by an accumulation of peat. They are dominated by a hummocky growth of *Sphagnum* covered by a shrub layer of heaths and later, in North America, by black spruce. Marshes, sedge meadows, and swamps occupy aerated soils subject to changing water levels, to spring flooding. They are often called bogs, but they accumulate little or no peat.

A *marsh* (fig. 1-4) is an open, grassy or sedgy wetland developed on mineral soil and standing under shallow water at least part of the year. The mineral content of the water is high, and the pH stands at or somewhat above neutrality. Productivity is high, but oxygen saturation favors decomposer activities. *Sedge meadows* are similar, but dryer than marshes during the heat of summer. Marshes and sedge meadows are succeeded by a wet shrub community of willows, dogwoods, and alders. Fens differ from sedge meadows in having a constant supply of water more or less rich in calcium and in accumulating peat.

A *swamp* (figs. 1-5–6) is a wooded wetland, rich in minerals and near-neutral to somewhat basic in reaction. It is highly productive. The oxygen saturation is high, owing to subsurface water movement, and peat development is limited by decomposer activities. The water movement may result from seasonal changes in water level as well as slope drainage. In the southern part of the Great Lakes region, hardwood swamps dominated by red maple, black ash, yellow birch, and (formerly) American elm are common (figs. 1-8–12), often in ice-block depressions and at the margins of kettlehole lakes. Pin oak swamps are also a feature of poorly drained soils (fig. 1-7). Tamarack swamps are common (often on glacial outwash and in ice-block depressions), as are shrubby swamps dominated by *Cephalanthus, Ilex, Decodon,* or *Cornus stolonifera* (figs. 3-47, 3-50, 3-52–53). Farther north hardwood swamps are largely replaced by tamarack and white cedar swamps. Cedar swamps favor a calcareous soil. They develop on a wet mineral soil or on lake sediments. Trees commonly associated with white cedar are balsam fir, tamarack, white spruce, black spruce, and yellow birch (figs. 1-14–17). (Black spruce does well in cedar swamps, on localized peat, but it does better in bogs. White spruce prefers a mineral substrate and does not occur in bogs.) Peat accumulation, though limited in cedar swamps, creates a degree of acidity at the surface. Wet depressions, including vernal pools, and rotting logs and stumps, as well as peaty turf, provide a diversity of habitats for herbs and mosses at ground level. A number of *Sphagna* are restricted to cedar swamps, but they by no means control the environment. Tamarack is a pioneer of considerable tolerance. It is succeeded by white cedar (or, in a bog series, by black spruce), but tamarack swamps sometimes occupy sites of less mineral enrichment than cedar swamps. Cedar and tamarack swamps are sometimes called *treed fens,* but they belong to a different hydrological regime from the open fens that lead to bogs. Generally speaking, fens do not develop into swamps unless there is a strong and continued flow of subsurface water. In Minnesota, the best development of a white cedar swamp occurs on slopes greater than 8 feet per mile (22). The drainage flow thus ensures a favorable nutrient supply and good aeration. In northern Michigan, *Thuja* swamps are usually associated with seepage water, but calcareous lakes filled with muddy sediments sometimes form a substrate suited to *Thuja* invasion. Rich fens, rich in calcium, become coniferous swamps. *Thuja* swamps are generally richer in base content, *Larix* swamps poorer. *Alnus* is commonly associated with *Thuja*, *Betula pumila* (fig. 1-13) with *Larix*.

The Varieties of Peatland Classifications

The kinds of wetlands can scarcely be compartmentalized and labeled. No two of them are exactly alike. The differences between marshes, sedge meadows, and fens can be quite intangible, and setback disturbances add to the problems of fen-bog classification. The stages of vegetational change merge together, and regional differences in species composition require consideration. Classifications devised for use in Europe are not entirely suitable on this side of the water; schemes useful next door in Ontario are not entirely applicable in northern Michigan; the large peatlands of north-central Minnesota are unlike those of more limited extent and different developmental histories elsewhere in the Great Lakes region; peatlands of oceanic areas are, to an extent, unlike those of the continental interior in origin, development, and floristics. However, some similarities provide a basis for a number of generalized classifications.

Peatlands can be classed, in a clumsy way, by water retention: *Primary peatlands* form within water-filled depressions or basins. In such peatlands, especially characterizing warm climates, in the Everglades, for example, evapotranspiration from emergent vegetation increases water loss to the atmosphere to the extent that no gain in water retention results. *Secondary peatlands* grow upward beyond the confines of the basin while remaining under the influence of ground or surface water, but the volume of spongy peat increases the amount of water retained. *Tertiary peatlands* grow beyond the level of the groundwater source of minerals because of a water table perched above an accumulation of impermeable peat. They retain water of precipitation. They are bogs, often called *raised bogs,* as contrasted with the flat secondary peatlands, or fens. In oceanic climates raised bogs may be elevated as much as 30 feet above the regional groundwater and conspicuously domed. Those of the continental interior show only a slight convexity. Raised they are, but not obviously domed. In the Upper Midwest, lake sediments may be considered primary peat, sedge deposits secondary, and *Sphagnum* deposits tertiary.

The classification based on water retention, as primary, secondary, and tertiary peatlands, is comparable to a number of others met with in the literature that express similar zonations, or stages of development, from sedge fen to poor fen to bog, while emphasizing different ways of explaining them.

Successional Development

Niedermoor	Uebergangsmoor	Hochmoor
Flachmoor	Zwischenmoor	Hochmoor
flat bog	transition bog	raised bog
low moor	transition moor	high moor

Phytosociology

| riekarr | karr | moss |

Nutrient Status and Floristics

rich fen	poor fen	moss
euminerotrophic mire	weakly minerotrophic mire	ombrotrophic mire
(eutrophic)	(mesotrophic)	(oligotrophic)

Hydrology, Chemistry, and Floristics

| rheophilous mire | transition mire | ombrotrophic mire |

Obvious differences in pH and mineral nutrition make it useful to think of bogs as *oligotrophic* (poor in nutrients) and *ombrotrophic* (deriving minerals solely from the atmosphere) and fens as *eutrophic* or *minerotrophic* (rich in nutrients) and *soligenous* (under the influence of mineral-rich ground or surface water). The best all-purpose classifications are based on water source and movement. Bogs develop under ombrotrophic conditions in which water and nutrients are supplied only from precipitation because water from the mineral soil is blocked by impermeable peat and cannot reach the surface layers. All earlier phases of development are minerotrophic fens in which water and nutrients are derived from above and also from below, in precipitation and groundwater.

Sjörs (230) recognized in the peatlands of northern Sweden a floristic continuum of six stages correlated with pH values and ion concentrations: *moss* (pH 3.7–4.2), *extreme poor fen* (pH 3.8–5.0), *transitional poor fen* (pH 4.8–5.7), *intermediate fen* (pH 5.2–6.4), *transitional rich fen* (pH 5.8–7.0), and *extreme rich fen* (pH ~7.0–~8.0). The overlap in pH values is even greater than the "normal values" given here, and the limits between the stages are especially blurred by an *"indeterminable poor fen"* for which Sjörs's notes were "insufficient to make it clear if a sample is from extreme or transitional poor fen." Sjörs considered the pH values and ionic concentrations significant only for northern Sweden and for determinations made by the same methods. Also, since his was basically a floristic classification, it cannot be applied with complete confidence in any other area, but when reduced to three groups, as *rich fen*, *poor fen*, and *moss* (or bog), it conforms well to differences in vegetation and chemistry readily observed in the Upper Midwest.

Bellamy (18) demonstrated seven peatland types based on hydrological, chemical, and floristic data taken throughout western Europe. These types can be grouped under three headings as *rheophilous mires* under the influence of groundwater derived outside the immediate catchment, with mean values of pH 5.6–7.5 and Ca^{++} and HCO_3^- the predominant ions; *transition mires* influenced by flowing water derived from the immediate catchment, with mean pH values of 4.1–4.8 and Ca^{++} and SO_4^{--} the major ions; and *ombrophilous mires*, never subject to flowing groundwater, with a mean pH of 3.8 and predominant ions H^+ and SO_4^{--}. These mires, or peatlands, are entirely comparable to those based on water retention (primary, secondary, and tertiary peatlands) and to Sjörs's peatland types (rich fen, poor fen, and moss).

In terms of vegetational zonation and fen-to-bog succession in the Upper Midwest, it is convenient to recognize in this treatment four peatland communities rather than three: *sedge fens* (intermediate fens—figs. 1-1, 2-9), *Sphagnum lawns* (poor fens), *open bogs* (hummock-hollow bogs—figs. 2-13–14), and *black spruce muskegs* (fig. 2-15).

The pH ranges are subject to seasonal variation, but a useful approximation is that bogs commonly have a pH of about 3–4, poor fens 4–6, and richer fens about 6–7.5. (This range of pH is equal to the difference between vinegar and sea water.) Indicator species and associations act as guides to conditions of water flow and chemistry in each of these vegetational types. The distinction between a rich fen that develops into a cedar swamp and a sedge fen that is less minerotrophic and goes into a bog line of succession is marked by a near-neutral pH and a *calcareous soil water limit* of about 18 mg/l (250). The change from fen to bog is marked by a *groundwater limit* and a pH of about 4 and a calcium content of approximately 1 mg/l (65, 279). Values may vary from place to place, and the dilution effect of precipitation may alter the limits. The calcium content is important in determining pH because calcium combines with carbonic acid, H_2CO_3, from the rain to form calcium bicarbonate, $Ca(HCO_3)_2$, which dissociates into Ca^{++} and HCO_3^- and further into H^+ and CO_3^{--} and buffers natural waters at pH values of 6–8. Nutrient availability and decomposer diversity are favored at such levels of pH. Some calcium content is characteristic of peatland water, regardless of its source and acidity. In Minnesota peatlands, for example, precipitation water typically contains 0.3–2 ppm of calcium, that provided by runoff at and near the surface has 2–10 ppm, and groundwater holds more than 10 ppm (and sometimes even 30 or more) (22). Bogs have calcium sufficient for plant growth, but they are poor in bicarbonate and therefore unbuffered and very acidic. Fens, bicarbonate buffered, have less acid or even alkaline pH values.

It is sometimes useful to classify fens and bogs purely by floristic differences as revealed by plant associations. Braun-Blanquet phytosociological designations are often used for peatland classification. A poor fen association dominated by *Carex limosa, C. livida,* and *Scheuchzeria palustris,* for example, can be named for the species of greatest constance a Caricetum limosae. Over its central European range this community has recently been divided into as few as eleven subassociations, but in the literature as many as 150 names exist for variant expressions of the same association, in fact, more names than species in the association (68)! Taxonomies based on species dominance can be relatively meaningless except in a localized context. In Newfoundland (285), the Canadian Wetland Classification scheme making use of Braun-Blanquet terminology has been used with some success. That scheme recognizes wetland *classes* (bog, fen, marsh, swamp, and open, shallow water), divided into *forms* based on topography (domed bog, plateau bog, ladder fen, etc.), then into *types* based on vegetational physiognomy (treed fens, wet

Sphagnum lawns, dry *Sphagnum* hummocks, etc.), and finally into *associations* (Kalmio-Sphagnetum fusci, Scirpo-Sphagnetum magellanici, etc.). This kind of hierarchical scheme can be used with different levels of precision depending on the user's needs.

Topographic variations (fig. 1-18) may be correlated with differences in climate, water source, and water movement. In the Upper Midwest and elsewhere in continental regions, flat fens develop into raised bogs of slight convexity. In oceanic regions, in parts of Newfoundland, the Canadian Maritime Provinces, and coastal Maine, as well as Great Britain and Baltic Europe, raised bogs may be exaggerated in elevation and convexity. (Some raised bogs of oceanic Britain exceed one mile in width and 15 or even 30 feet in height.) The centers are nearly flat and the marginal slopes gentle. Near the coast may be *plateau raised bogs* with sloping, often wooded margins and relatively flat tops with an unpatterned arrangement of hummocks and hollows or pools (figs. 1-19–21). Somewhat inland the bogs may be conspicuously domed. Those originating on flat substrates are *concentric domed* (figs. 1-23–24), with elongated hummocks and hollows (or pools) perpendicular to the slope. On a sloping terrain the bogs are lopsided, or *eccentric domed* (fig. 1-25), with hummocks and hollows more abundant on the longer slope. A domed bog characteristically has an inclined margin (or *rand*) and a surrounding *lagg*, into which water drains from the dome and from surrounding slopes. Where the lagg drains at the downslope end is an *apron* which, like the lagg, supports a sedgy or shrubby growth as a response to more or less mineral-rich water.

In the course of succession, a peatland eventually comes to stand above the level where groundwater can be lifted upward by capillarity. Above that level, atmospheric water forms a perched water table. At this point, the peatland becomes an ombrotrophic mire, a bog, rather than a soligenous fen. The formation of a dome depends on unbalanced growth and decomposition. In the center of the bog *Sphagnum* grows faster than it can be decayed away, but at the margins mineral-enriched and aerated water causes decomposition to exceed productivity in and along the lagg. An abrupt imbalance in peat accumulation results in doming above an abruptly inclined slope. The slope may be steeper along the downslope margins because of vegetational responses in productivity to minerals in the lagg. The domed surface follows the convexity of the perched water table at its summer low because of gravitational and frictional restraints on capillary water movement.

Along the coast, in wetter parts of Newfoundland, Scotland, England, Wales, and Ireland, for example, peatlands blanket vast expanses of irregular terrain, even on slopes of 10° or 25°. *Blanket bogs* (figs. 1-26–27) are encouraged in their development by the formation of impermeable hardpans in the subsoil. *Sphagnum* is a relatively inconspicuous feature of such blanket bogs, which have a drab appearance (generally brown) owing to a prominence of mosses other than *Sphagnum*, including *Rhacomitrium lanuginosum*, lichens, and above all, dead remains of such sedges as *Rhynchospora, Scirpus*, and *Eriophorum*. Such heath plants as *Kalmia* and *Empetrum* impart a dark red monotone in winter. Blanket bogs are more oceanic than domed bogs. They derive minerals solely from the air and are relatively base deficient, although salt sprays contribute to their mineral nutrition. Water moves through them, at least superficially. For that reason a blanket bog might well be considered an acid fen in which downslope drainage introduces a degree of aeration and peat degradation. In western Ireland, 60 inches of rainfall favor the formation of blanket bogs, near the coast. Inland, in the Irish Midlands, 40 inches favor the vast raised bogs now being exploited—yes, depleted—by mechanized peat winning on a large scale. The transition from blanket bog to raised bog is marked at an annual precipitation of about 40 inches and the change from raised bog to fen at about 20 inches (152). (It is, of course, the persistence of humidity and the precipitation:evaporation ratio that make the difference, not the actual amount of precipitation.)

Slope bogs (fig. 1-22) are similar to blanket bogs but limited in extent by topography. Usually small, they probably originate as a result of infilling at the base of a slope followed by waterlogging and upslope invasion by peatland vegetation. Blanket bogs are more extensive, covering depressions as well as slopes in irregular terrain. (The distinction between slope and blanket bogs is often difficult to make.)

Well inland, in southern Finland and the Hudson Bay lowlands, for example, extensive patterned peatlands, or *aapamires* (fig. 1-28), may form on gentle, often imperceptible slopes. The flow of water is often marked by a ladderlike arrangement of long ridges, or *strings*, alternating with wet depressions, or *flarks*, lying perpendicular to the slope (fig. 1-29). Patterned peatlands are a special feature of northern continental climates. They occur in a band across Canada and Alaska, south of the zone of discontinuous permafrost. There is little sign of doming owing to a limited production of tertiary peat. Because of moving water, both strings and flarks are fenlike, although some approach to bog conditions can be found on well-formed strings (fig. 1-30). It is difficult to separate string bogs from fens in terms of water chemistry and peat thickness, except that the strings may eventually become built up enough to support a boglike growth of *Larix*, *Picea*, and ericaceous shrubs. North American aapamires commonly have stands of black spruce on islandlike mounds tapered downslope (fig. 1-31). Irregularly patterned mires are found in Michigan's Upper Peninsula, in Alger, Luce, Mackinac, and Schoolcraft counties (102, 140) (figs. 1-33–34). Those of the Red Lake peatlands of northern Minnesota (fig. 1-32) are more obviously patterned, some ladderlike and others reticulate. Patterned fens are also found in Maine (239, 281) and very sparsely in New York and Wisconsin.

Aapamire formation is certainly related to water flow. On a slope, even a very slight one, both strings and flarks are narrowly oblong or linear and their elongation perpendicular to the slope is a consequence of the fact that the water level is exactly the same at any point on the same contour. But the slope is so slight, sometimes as little as 1°, that water flow can scarcely provide the sole explanation. The patterns have been attributed to peat sliding downslope, greater decay in pools owing to production of oxygen by algal growth, ice pressure in pools pushing up ridges, differential thawing, plant material deposited in windrows by runoff, runner plants forming clones perpendicular to slopes, and substrate topography. There is probably no one explanation, but the following one has merit (71, 87, 88). After formation of an irregular, hummocky topography, occasional flooding may slow down peat formation in hollows, but continued growth of hummocks raises the water table and joins water-filled hollows at the same level crosswise of the slope, giving rise to elongated flarks alternating with ever-growing and joining strings. Reticulate patterns may be related to drainage pressures breaking strings and joining flarks. (As recently suggested by Barbara Madsen [140], the most basic reason for the actual origin of strings and flarks may be the seepage force of water moving over sloping terrain, causing the compression of peat in some places and the cracking at other places under strain. Freezing and thawing and oxidative degradation may accentuate the features after their formation.)

Permafrosted Peatlands

In the far north, where forest turns to tundra, drainage may be impeded by patches of permafrost. Southward, in a zone of discontinuous permafrost, isolated mounds, or *palsas* (fig. 1-35), may develop to a height of 10 or 15 feet and to a diameter as much as 410 feet. The palsas are more numerous northward, closer to the zone of continuous permafrost, and as they grow, they coalesce as *peat ridges* and *peat plateaus* (figs. 1-37–38) that give *palsa mires* reticulate or marbloid *airform patterns* as seen from high altitudes (fig. 1-40). The palsas originate as peat upwarped from shallow pools that freeze to the bottom in winter. Continued elevation results from ice layers forming beneath the warp. Dry peat at the surface and a developing cover of vegetation prevent thawing in the summer, but exposure to winter winds reduces the insulating effect of snow cover and allows additional freezing. Water drawn in from wet surroundings in the summer and frozen in the winter contributes to a continued frost accumulation at the core, year after year. The palsas are covered by mosses and lichens and eventually an overstory of Labrador tea. In eastern Canada they also support a stunted growth of tamarack and black spruce (fig. 1-39), but those formed west of Hudson Bay, like those of Finland, are north of the tree line. The ground vegetation is eventually shaded out, and the peat, exposed to drying winds, erodes and decays. The resulting penetration of summer heat causes the ice core to melt and the mound to collapse, sometimes leaving behind a ringlike stand of spruce. Sometimes a *drunken forest* of leaning trees results from a melting core. The collapsed central portion commonly supports a sedgy fen growth and is often referred to as a *collapse scar* (fig. 1-36).

In the Canadian subarctic, south of the palsa zone, are frozen mounds of another kind. Such mounds, or *black spruce islands* (figs. 1-31–32), occur in series along small streams and also scattered in open fens. In scattered form, they are characteristically shaped like teardrops tapered downslope. The tapered tail is only slightly elevated and supports a shrubby vegetation. It has been suggested that these mounds developed like palsas at some period of colder climate or that black spruce layering allowed a continued accumulation of winter ice during climatic conditions that now prevail. Whatever the explanation, the spruce islands far to the south in Minnesota (fig. 1-32) can scarcely be attributed to colder climates of the past, as stratigraphic studies reveal no evidence of much cooling since the origin of those peatlands.

In the arctic (and especially in the Mackenzie River delta), huge mounds, as much as 250 feet high, form in lakes as a result of hydraulic pressure. These *pingos* result from the upwelling of sediments through cracks in the permafrost and continued pressures caused by the lateral expansion of permafrost in lake-fill sediments. Pingos are covered with alluvium rather than peat.

Where the drainage is impeded by continuous permafrost, peatlands may be more or less continuous or small and scattered. In bare soil regions of the arctic, the landscapes are patterned by the sorting action of frost churning and the formation of ice wedges delimiting large polygonal areas (fig. 1-41). In depressed areas, in ditches surrounding polygons, for example, an insulating cover of snow prevents frost penetration, and drainage water collects there. Mires develop in the wetness surrounding ice-wedge polygons and eventually grow up over the terrain to form discontinuous *terrazoid* airform patterns of peatland cover and, in time, more continuous marbloid patterns.

In summary, as a basis for understanding peatlands in the Upper Midwest, a developmental classification is used. Rich fens become cedar swamps in an entirely minerotrophic series, but the bog series begins with minerotrophy and ends with ombrotrophy. The bog line of development begins with intermediate fens (with a lesser calcium content than rich fens), followed by poor fens (*Sphagnum* lawns), open bogs (low shrub, hummock-hollow complexes), and finally black spruce muskegs. The rich fen–*Thuja* series is outside the developmental scheme of wetlands that continue to accumulate peat and therefore become progressively higher, dryer, and more acid. With a totally different hydrology, it is totally different at maturity from a boggy muskeg.

Fig. 1-1. An intermediate fen, rich in minerals but less calcareous than a rich fen, at the edge of Inverness Mud Lake, Cheboygan County, Michigan. The fen, dominated by *Carex lasiocarpa*, is typical of the mat that forms at the edge of an alkaline lake with a false bottom of sediments. The cattails scattered in the mat are associated with water movement caused by a nearby inlet.

Fig. 1-2. Livingston Bog, Cheboygan County, Michigan, an acid, kettlehole bog with a pioneering *Chamaedaphne* and *Andromeda* mat and a narrow *Sphagnum* lawn behind it. (Photo from the files of the University of Michigan Biological Station.)

Fig. 1-3. Stutsmanville Bog, Emmet County, Michigan, surrounding an acid lake with a narrow margin of *Chamaedaphne* growing out into the shallow water and, in places, a narrow mat of *Carex limosa* and *Rhynchospora alba* instead. The pioneering mat is followed by a narrow *Sphagnum* lawn (or poor fen) abruptly succeeded by tamarack and black spruce with an understory of *Ledum*.

Fig. 1-4. Marsh at inlet to Crooked Lake, Emmet County, Michigan. A marsh is associated with a high water table and moving water that is aerated and conducive to microbial activity such that decomposition equals productivity. The mineral-rich water is indicated by floating-leaved macrophytes and a "grassy" vegetation that includes cattails.

Fig. 1-5. Rich fens and cedar swamps developed from them are associated with a constant flow subsurface water rich in calcium

Fig. 1-6. The shallow pools in open rich fens commonly dry up during the summer exposing a marly "pond porridge" of obvious lime content.

Figs. 1-5–6. Cedar-tamarack swamp developed in a rich fen at Summerby Swamp, Mackinac County, Michigan

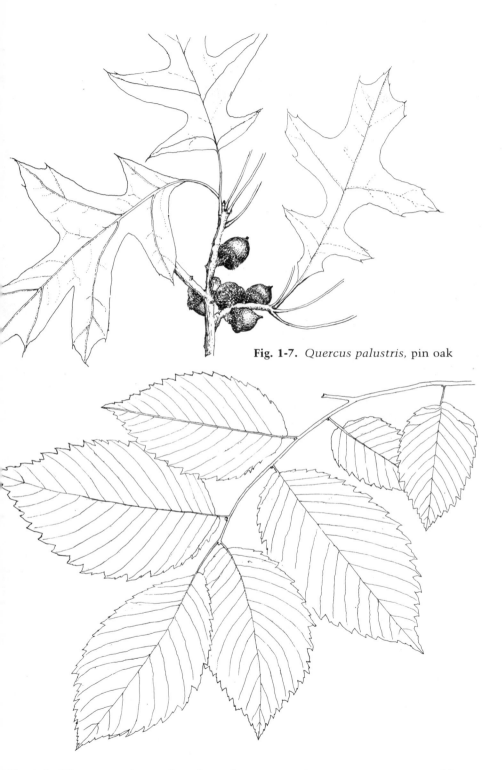

Fig. 1-7. *Quercus palustris*, pin oak

Fig. 1-8. *Ulmus americana*, American elm

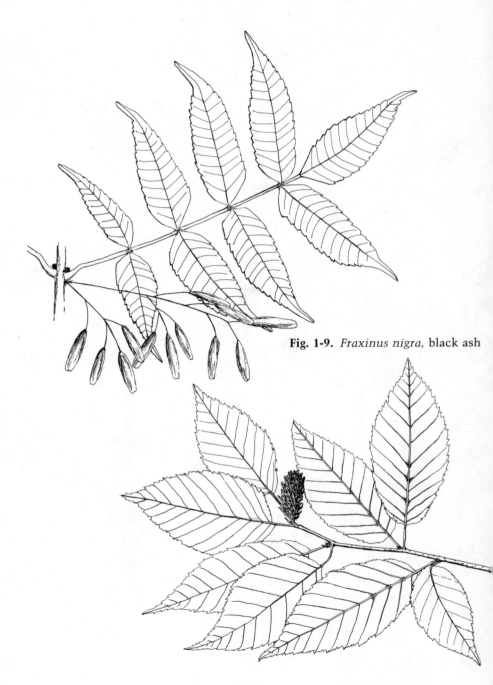

Fig. 1-9. *Fraxinus nigra*, black ash

Fig. 1-10. *Betula alleghaniensis*, yellow birch, common in *Thuja* swamps

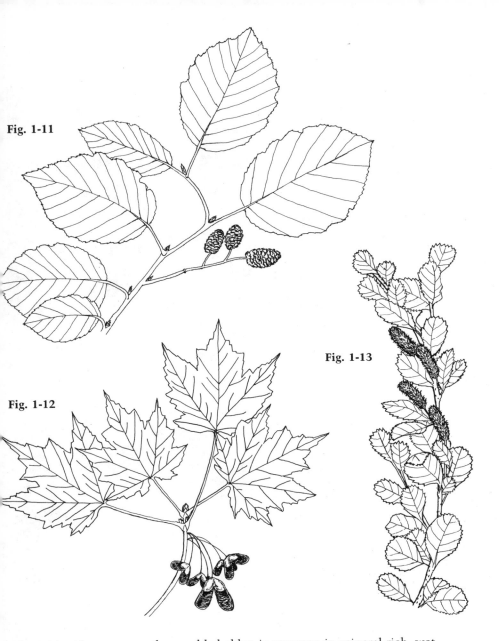

Fig. 1-11. *Alnus rugosa*, the speckled alder, is common in mineral-rich, wet habitats, including especially shrubby swamps and laggs surrounding lake-fill peatland. **Fig. 1-12.** *Acer rubrum*, red maple, is very common in the wooded carrs that form on older parts of sedge mats in the southern part of the Great Lakes region. It grows in hardwood swamps in the south and also far to the north, and it also pioneers dry upland soils. **Fig. 1-13.** *Betula pumila*, dwarf or bog birch, is a low shrub of mineral-rich wetlands in the North. It may be found in shrubby swamps formed along waterways or in association with tamarack in poor fens transitional to bogs.

Fig. 1-14. *Thuja occidentalis*, northern white cedar, is the dominant tree of cedar swamps.

Fig. 1-15. *Abies balsamea*, balsam fir, is a tree of cedar swamps occupying somewhat drier sites than *Thuja*.

Fig. 1-16. *Picea mariana*, black spruce, occurs in *Thuja* swamps but is more characteristic of older parts of bogs.

Fig. 1-17. *Larix laricina*, tamarack, a deciduous-leafed conifer of relatively mineral-rich places, occurs in *Thuja* or tamarack swamps and often grows on hummocks at transitions between floating mat and grounded bog.

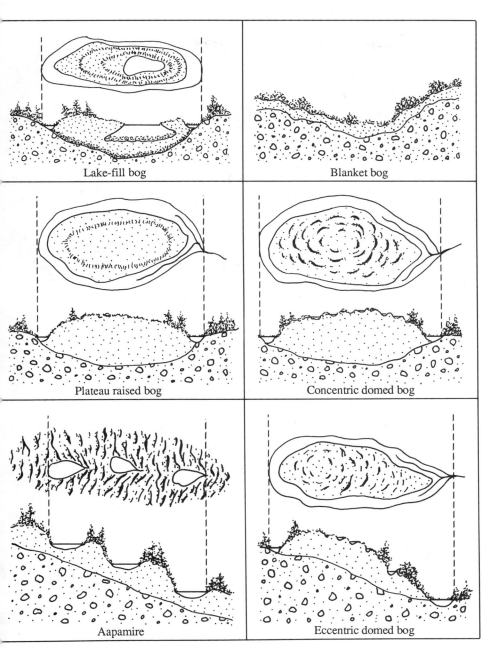

Fig. 1-18. A schematic representation of peatlands categorized by surface topography (the topographic relief is greatly exaggerated)

Fig. 1-19. A plateau raised bog complex in Newfoundland. The numerous pools are scattered over the flatness of the peatland, whereas in a concentric or eccentric domed bog they are arranged in a ladderlike fashion perpendicular to the bog slope. (Photo by Doyle Wells.)

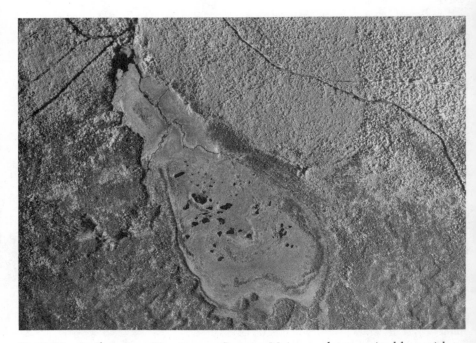

Fig. 1-20. Caribou Bog, Piscataquas County, Maine, a plateau raised bog with a well-developed apron where lagg waters discharge. (Photo provided by Harry Tyler.)

Fig. 1-21. Plateau raised bog surrounded by a well-marked lagg, Penobscot County, Maine. (Photo provided by Harry Tyler.)

Fig. 1-22. Slope bog, Washington County, Maine. (Photo provided by Harry Tyler.)

Fig. 1-23. A concentric domed bog in Newfoundland. (Photo by Doyle Wells.)

Fig. 1-24. A domed bog in Newfoundland with a nearly concentric arrangement of pools. (Photo by Doyle Wells and F. C. Pollett.)

Fig. 1-25. An eccentric domed bog in Aroostook County, Maine. (Photo provided by Harry Tyler.)

Fig. 1-26. Newfoundland blanket bog on sloping terrain in the Avalon Peninsula

Fig. 1-27. The surface of a Newfoundland blanket-bog, with patches of snow in mid-April and windswept tufts of *Scirpus cespitosus* scattered among low mounds of *Sphagnum* topped by the moss *Rhacomitrium lanuginosum* and reindeer lichens of the genus *Cladina*. The winter aspect is a dull red-brown expanse colored by persistent leaves of *Kalmia* and *Empetrum*.

Fig. 1-28. Aapamire (ribbed fen) in the Hudson Bay Lowlands of northeastern Manitoba, showing a ladderlike arrangement of strings and flarks. (Photo by Lynda Dredge.)

Fig. 1-29. A string covered with a poor growth of tamarack and a sedge fen in the flark behind it. The flarks are often filled with water rather than supporting a sedgy fen growth. A black spruce muskeg is in the background. (Photo by Lynda Dredge.)

Fig. 1-30. International Fen, Somerset County, Maine, showing water-filled flark bordered by strings. (Photo provided by Harry Tyler.)

Fig. 1-31. Tear-shaped islands in an aapamire in northeastern Manitoba. (Photo by Lynda Dredge.)

Fig. 1-32. Long-tapered islands of vegetation oriented perpendicular to strings and flarks in northern Minnesota. As in fig. 1-31, the tapered ends of the islands are directed downslope. (Photo provided by Frank Bowers.)

Fig. 1-33. Patterned peatland, Creighton Marsh, Schoolcraft County, Michigan. (Photo by Barbara Madsen.)

Fig. 1-34. Patterned peatland west of St. Ignace, Mackinac County, Michigan. (Photo by William Taylor.)

Fig. 1-35

Fig. 1-36

Figs. 1-35–36. Small palsas developed in a wet, open fen. The palsa below has collapsed owing to the melting of its ice core. The depression so formed is known as a collapse scar. (Photos by Lynda Dredge.)

Fig. 1-37. Peat plateau in the Hudson Bay Lowlands of northeastern Manitoba, presumably developed from a palsa with an ever-growing core of ice. (Photo by Lynda Dredge.)

Fig. 1-38. Peat plateau with an accumulation of peat 2 meters in depth. (Photo by Lynda Dredge.)

Fig. 1-39. Muskeg in the of discontinuous permafr the Hudson Bay Lowland northeastern Manitoba, showing black spruce wi characteristic bunching c branches at top. (Photo b Lynda Dredge.)

Fig. 1-40. Coalesced peat plateaus in the Hudson Bay Lowlands of northeastern Manitoba. The plateaus occupy areas of permafrost and are probably overgrown palsas. (Photo by Lynda Dredge.)

Fig. 1-41. Ice-wedge polygons associated with frost churning and upheaval, Hudson Bay Lowlands, northeastern Manitoba. Embryo mires develop in the drainage channels separating polygons. Such mires may come to cover the entire surface of the polygon. (Photo by Lynda Dredge.)

Chapter 2

The Origin and Development of Peatlands

> All change is a miracle to contemplate; but it is a miracle which is taking place every instant.
> —Thoreau

Peatlands are particularly well developed in oceanic areas with long growing seasons, with cool, humid, foggy summers and mild winters. Extreme cold and drought make the continental interior less suitable, but in the Great Lakes states, in northern Minnesota, Wisconsin, and Michigan, poor drainage associated with glacial topography and an even distribution of rainfall and atmospheric humidity make for peat accumulation, even though the summers are hot and the winters long and cold. Mean annual precipitations are 20 inches in northwestern Minnesota and 34 inches in Michigan's Upper Peninsula, and about two-thirds of the precipitation takes place from April to September. At Douglas Lake, in the northern part of the Lower Peninsula, precipitation is only about 30 inches a year, in contrast to 50 inches in oceanic Nova Scotia, but it is distributed throughout the summer, even in the heat of August. In the southern Lower Peninsula, the amount of precipitation is similar to that in the north, but August is hot and dry. The result is that fens develop in southern Michigan, but the greater accumulation of peat and continued rainfall conducive to bog formation are lacking. The boreal and subarctic regions of the continent support a virtually unbroken expanse of peatland from the Atlantic to the Pacific (figs. 2-1–3). Yet, in terms of precipitation, the far north is a desert. It is poor drainage coupled with reduced summer temperatures that makes peat accumulation possible there. The poor drainage results from the precipitation-evaporation balance, as well as topographic uniformity and impermeability of clay, peat, and permafrosted soil.

Peat-storing ecosystems form where the water table stands at or near the surface, as a result of *lake-fill* (figs. 2-4–8) or *paludification*, also known as swamping (figs. 2-7, 2-30). They form at the edge of lakes in a well-zoned fen-to-bog forest sequence or in wet depressions where zonation, if any, is less evident. In the Upper Midwest the landforms conducive to peatland formation are those shaped by glacial retreat or postglacial wind and wave action. Lake-fill basins encroached on and eventually covered by mats of vegetation sometimes have their origin in kettleholes once occupied by ice blocks buried in glacial outwash. Kettlehole lakes are characterized by an abrupt drop-off and cold water. As small lakes surrounded by woodlands, they are protected, to an extent, from wind and wave action and seasonal overturn. After glacial retreat, great areas were left as meltwater lakes where the reworking of bottom soils by wave action often pushed up bars and spits to delimit quiet-water embayments where lake-fill processes could get a start. Series of dunes piled up by onshore winds provided depressions where water levels suited peatland initiation. Rising water tables associated with postglacial climatic changes also created lakes in lowlands or provided soggy wet conditions for paludification.

In the case of lake-fill, peat is deposited over lake sediments. But in paludified areas, peat is deposited on land that for one reason or another became wet, as indicated by remnants of forest vegetation commonly buried below the peat. It is also possible for peatlands to develop on newly available soils associated with glacial retreat, river delta sediments, and coastal uplift. Whatever the origin, conditions of the habitat accommodate a similar vegetation and a similar fen-to-bog succession.

In most ecosystems the mineral soil dictates, to a large extent, the composition of the vegetation, but in peatlands the soil is a product of the vegetation, and in ombrotrophic peatlands the ground surface is most commonly living *Sphagnum*. The *Sphagnum*-dominated vegetation controls the ecosystem. By producing peat, it alters the quantity and quality of water flow and eliminates essential nutrients from cycling. The development of a peatland means the eventual depletion of nutrients, stored away in peat, unavailable for circulation. The changes in hydrology and nutrition affect the selection of plants at each stage of succession, and at each stage the community both destroys and transforms itself. By creating and maintaining an acid, waterlogged habitat, *Sphagnum* selects against competitors for light: It encourages a loose growth of low shrubs that can continually produce new roots and shoots above oxygen-poor wetness, and it lays down a soil unsuited to the support and nutrition of trees.

As the vegetation shifts toward maturity, the species diversity becomes progressively less because of poor nutrition and habitat uniformity. The vegetation has a different physiognomy at each stage of succession. Vegetational changes follow gradients in water and nutrient supply, and successional zonation is expressed by the kinds of plants, their relative abundance, and their coexistence in communities. The plant communities characterizing each stage show differences associated with regional climates, montane and lowland, oceanic and continental, northern and southern. Because of different ecological amplitudes, species that grow together at one part of the range may occupy quite different habitats elsewhere. Thus, *Scirpus cespitosus* is an acidophile in coastal Newfoundland but a calciphile in the Upper Midwest. *Carex limosa, Scheuchzeria palustris,* and *Sphagnum majus* are restricted to fens in southern Sweden but also grow, facultatively, in bogs farther north. *Eriophorum vaginatum,* on the other hand, is restricted to bogs and poor fens in the south of Sweden but grows equally well in extremely rich fens in the north. *Sphagnum papillosum, S. imbricatum, S. magellanicum,* and *S. rubellum* grow in poor fens in eastern Sweden, but in the southwestern part of that country they occupy acid bogs. *Schoenus* and *Molinia* are restricted to fens in the continental interior, but in Ireland they characterize distinctly ombrotrophic blanket bogs. The high rainfall in the Irish habitats causes an almost continuous runoff at the surface, and the high mobility of water makes the blanket bogs fenlike in oxygen richness. Rainwater carries mineral ions, of course, and continual flushing may remove hydrogen and sulfate ions from solution and lessen the acidity, and ions brought in by ocean spray aid in removing acid ions owing to chemical combinations. The oxygen associated with moving water increases the rate of decomposition and the amount of nutrients recycled, and also flowing water increases nutrient availability if only because more ions are supplied per unit time.

Extremes in wetness, oxygen supply, and nutrient level, as in temperature and acidity, limit the plants that can survive the bog environment. The species are similar or identical over much of the Northern Hemisphere, although varying in associations. Endemics and rare disjuncts are not to be expected in the wide-ranging uniformity of fen and bog habitats. The assemblage of species in rich, calcareous fens is not at all like that in acid bogs, but habitat extremes shade off into other kinds of peatlands. Many rich fen species occur in the intermediate fens bordering mineral-rich lakes, and some of them in turn grow in poor fens. The vascular plants have a "high trophic plasticity." Mosses are better habitat indicators, but the associations of vascular plant species and their relative abundances, as well as the physiognomy of vegetational communities, give an index to abiotic conditions.

Lake-Fill

> It was through one of the open bogs of Poland that a German officer ordered his men to attack the retreating Russians when practically the whole regiment sank out of sight. The officer himself escaped but became mentally unbalanced as a result of the catastrophe.
> —Hotson

Whatever their origins, by lake-fill or paludification, peatlands accumulate water-soaked organic matter that profoundly alters the habitat in chemistry and hydrology. Wet and compacted peat, virtually impervious to water, reduces the nutrient supply by sealing off the substrate, impeding the flow of groundwater, and blocking runoff from upland slopes. A reduced supply of nutrients shifts the chemical environment toward acidity even as *Sphagnum* species create acid niches by cation exchange.

Fens (fig. 2-9) pioneer lowlands of constant water supply and often encroach on open water. Developing under the influence of aerated ground or surface water, fens are open, sedgy habitats, rich in minerals and species, high in productivity. High concentrations of calcium bicarbonate buffer the pH at circumneutral values. Oxygen concentrations are at first relatively high, but waterlogged peat eventually impedes the movement of dissolved oxygen and diminishes the rate of decomposition. The buildup of anaerobic peat slowly transforms a sedgy wet fen through a wet *Sphagnum* lawn to a higher, dryer, acid bog complex dominated by *Sphagnum*. The hummock-hollow *Sphagnum* growth supports a cover of low-growing shrubs, particularly *Chamaedaphne*. Eventually the low-shrub complex is succeeded by black spruce. Bogs, being above the influence of groundwater, derive nutrients solely from precipitation and atmospheric dust (except for some possibility of springtime runoff from surrounding slopes). They are poor in nutrients and deficient in oxygen. The peat soil is cold and definitely acid. Species diversity is low, and so is productivity.

Fens and bogs may develop concentrically all around the margins of land-locked lakes, but lakes with outlets commonly form mats at the end opposite the drainage. The nature of the mat differs, especially in its pioneer stages, with the nature of the lake water, whether alkaline (or at least mineral rich) or acid.

"Alkaline" Lake Mats

The vegetation of a fen developed next to a mineral-rich body of water (buffered at or slightly above neutrality) consists largely of sedges, most commonly *Carex lasiocarpa*. Interwoven rhizomes form a *quaking mat* that floats up and down with changing lake levels. In this way, the more pioneering species of *Sphagnum*, so important in establishing a chemical environment suited to other plant invaders, are protected from submergence and from an excess of minerals, especially calcium and magnesium, to which they have only limited tolerance. The pioneer *Sphagnum* is commonly *S. teres* and sometimes, in especially calcareous, rich fen situations, *S. subsecundum* and *S. warnstorfii*.

A mineral-rich lake teems with algae and invertebrates living on them. The algae extract carbon dioxide from an abundant source of calcium bicarbonate for use in photosynthesis. As a result, calcium carbonate precipitates as *marl*, together with a compact, dark, gelatinous *gyttja* derived from planktonic sediments and a bottom fauna with its feces. Under conditions of nutrient excess, the algae and other respiring organisms (including decay bacteria) deplete the dissolved oxygen faster than it can be replaced by photosynthesis, and oxygen deficiency inhibits decomposition. The continual addition of the dead remains of countless small, even microscopic plants and animals produces a muddy suspension that nearly fills the basin as a false bottom. The false bottom eventually develops enough substance that aquatic macrophytes can be anchored in it, at least at the shallow periphery of the lake. These macrophytes include such indicators of eutrophy as water lilies and pondweeds. During dry years, when lake levels go down even a few inches, the false bottom may be exposed at the lake margins and sometimes much more extensively. It is then that the pioneering sedge mat moves outward (figs. 2-10–12).

In company with *Carex lasiocarpa*, such plants as *Menyanthes*, *Dulichium*, and *Potentilla palustris* are commonly represented in the advancing sedge mat. In this area of considerable species diversity, some few true mosses (as opposed to peat mosses) occur in the wetness hidden away at the base of sedges. Alders and cattails invade the mat where there is a subsurface inflow. Winds and waves acting on the edge of the mat, especially at the time of ice break, may cause upwarped "ice push ridges" that become occupied by tamarack and spruce. Sometimes islands of vegetation break off, float about, and become joined to some other part of the mat.

The sedge mat is succeeded by an open, poor fen stage in which *Sphagnum* forms a wet lawn. Sedges scattered in the *Sphagnum* lawn include species of *Rhynchospora*, *Carex*, and *Eriophorum*. Also scattered in the lawn may be *Scheuchzeria*, several orchids (*Pogonia*, *Calopogon*, *Arethusa*, and *Habenaria clavellata*), and the carnivores *Sarracenia* and *Drosera rotundifolia*. The cranberries and low ericad shrubs, *Andromeda* and *Kalmia polifolia*, occur there in low density. Hummocks get a start in the older parts of this zone and become a conspicuous feature later at the bog phase of development. *Chamaedaphne* may be found there, especially in the dryer, more mature parts of the lawn community, often with *Larix*.

Both the sedge mat and the *Sphagnum* lawn float, but the bog is grounded on firm peat. Tamarack, indicative of a groundwater influence, grows on scattered hummocks near the *hinge line*, in a zone of transition between fen and bog.

Plant diversity is exceedingly limited in the hummock-hollow complex of the bog. *Sphagnum capillifolium* and *S. magellanicum*, often mixed with *S. recurvum*, occupy the lower and intermediate levels of hummocks, and *S. fuscum* forms a compact, dry cap. In wet hollows formed by hummock degradation, and especially under clumps of black spruce, are a few bog liverworts. *Chamaedaphne* nearly covers the hummock-hollow topography of the open bog (fig. 2-13). Because its spreading rootstocks are protected by wet *Sphagnum*, *Chamaedaphne* survives superficial burns, but more severe fires favor an ingress of blueberries. (Indians and white settlers used to burn bogs to encourage blueberries.) More severe fires, destroying the hummocks in which trees and shrubs come to be rooted, cause a reversion to a wet sedge fen or even to standing water.

Black spruce replaces tamarack on the *grounded mat*. It grows first in scattered clumps in the low-shrub zone (fig. 2-14). The clumps represent clones. The spruce gets established on the hummocks. As a hummock grows to a height where it is too dry to be replenished by further *Sphagnum* growth, it deteriorates and becomes a hollow. Meanwhile, as the tree grows, it sinks into unstable peat. As its lower branches contact the wet *Sphagnum* of surrounding hummocks, they sprout new plants. Before it dies by drowning, the tree leaves a circle of offspring. Older parts of the bog, at the outer margin of the mat, support an open forest of black spruce, or *muskeg* (fig. 2-15). *Sphagnum* continues to be abundant in this treed zone, but *Chamaedaphne* is replaced by the more shade-tolerant *Ledum*. The forest floor provides habitats for feather mosses and lichens, as well as a scattering of herbs. The bog surface has at least a slight convexity, and the slope drainage has some significance in the establishment of a black spruce stand. The relative dryness of the hummocky terrain and drainage from the convexity result in greater aeration.

At the outer edge of the mat, beyond the spruce muskeg, is an open lagg, or moat (figs. 2-16–17), or a high-bush zone under the control of drainage water of mineral content. It may be relatively dry, or it may be wet, owing to runoff from surrounding slopes. Excess shade, aeration, mineral richness, seasonal changes in wetness, and considerable leaf litter limit peat-producing vegetation in the lagg.

Acid Lake Mats

In the case of a mineral-poor lake, the water is clear and the bottom firm. The microflora and microfauna are less abundant and aquatic macrophytes fewer. However, *Utricularia* submerged in shallow water may give sign of minerals secondarily derived by trapping invertebrates. *Chamaedaphne* often hangs out over the edge of a narrow mat (figs. 2-18–19), sometimes curling over and ending upside down under water, or it may be rooted under shallow water. A sparse fringe of *Carex* may also grow in the water. Both *Carex* and

Chamaedaphne give support for other plants, including *Sphagnum*, to become established, but *Sphagnum* alone often pioneers the mat. The narrow *Sphagnum* lawn abruptly gives way to a grounded mat occupied at its outer margin by *Chamaedaphne*, which is crowded into by tamarack, black spruce, and an understory of *Ledum*. Tamarack is a minor feature of this telescoped transition from poor fen to muskeg. It soon drops out because of intolerance to shade and reduced availability of nutrients.

Around a small lake, the fill-in can be remarkably fast (fig. 2-20), but larger lakes, exposed to more wind sweep, probably are much slower in mat formation.

Rich Fen Mats

At the margin of highly calcareous lakes, a rich fen may occupy exposed false bottom sediments and in time be succeeded by tamarack and white cedar. Such a mat can be treacherous underfoot, as the muddy false bottom of little structure can be covered with only a sparse growth of sedges and other plants suited to a pioneering habitat (fig. 2-22). *Carex lasiocarpa* eventually adds firmness to a mat that may also have as pioneers such calcium-tolerant mosses as *Sphagnum teres, S. subsecundum, S. warnstorfii, Scorpidium scorpioides, Calliergon trifarium, Calliergonella cuspidata,* and *Tomenthypnum nitens*, as well as the liverwort *Moerckia hibernica*.

Minerotrophic Lake Mats of the Southern Great Lakes Region

In hotter, dryer parts of the Upper Midwest, marshes commonly develop in the shallows of lakes, but peat-forming fens are more limited in occurrence. Peatlands may form as lake-mat fens dominated by cattails or grasses and sedges, but they fail to go into a bog stage of buildup. The fens give way to tamarack, poison sumac, red maple, and black ash. *Chamaedaphne* may be present, and *Sphagnum* is a significant component of such fens and fen *carrs* (that is, peatlands supporting the growth of deciduous-leaved trees and shrubs), but hummocks are poorly marked. Both horizontal and vertical successions from mineral-rich to mineral-poor and wet to dry are truncated. Successions beginning with *Sphagnum* lawn communities can also be found, though less commonly, as evidence of mineral-poor wetness. In such relatively acid mats, almost boglike, tamaracks thrive, but black spruce is represented in token amount, or not at all.

In southern Michigan, in wet depressions surrounded by the alkaline soils of end moraines, the fens appear to be relatively stable. Even though muck deposits may be as much as 73 feet deep, there is little evidence of a shift toward an acid bog replacement. It is probable that the distribution of rainfall and the precipitation-evaporation balance have not been suitable for further change.

Paludification

The vegetational zonation from open water to spruce muskeg is not entirely representative of peatlands or indicative of their developmental history. In North America, as in Europe, peat cores often show forest remains below bog deposits. In fact, the sequence is commonly aquatic, sedge, forest, and finally *Sphagnum* peat. In all regions where peatlands are really extensive, both in cool, oceanic regions and inland at north temperate to subarctic latitudes, paludification has been far more important than lake-fill in peatland initiation. The paludification of forests and the subsequent invasion by bog vegetation can often be attributed to shifts in climate. Postglacial climates were at first cold and wet, but about 7,600 years ago a warming trend set in. Warm, dry climates persisted until about 2,500–3,000 years ago when they were replaced by the relatively cool and moist conditions that have continued until now. The record of past climates, as evidenced by vegetation types, and, to an extent, human activities are preserved in lake muds and peat deposits. The chronology of vegetational responses to climatic change can be read through the nature of the peat and its pollen content. Brightly colored layers of relatively raw *Sphagnum* peat document cool, moist climates, but underlying dark, humic peat poor in recognizable plant structures records an advanced degree of decomposition associated with warmer, dryer climates and early stages in succession. In northwestern Europe and the British Isles, an abrupt change in the kind of peat radiocarbon-dated at 500–600 B.C. marks the beginning of a colder and rainier climate especially conducive to peat formation and also a shift from Bronze Age to Iron Age cultures. Pollen analysis shows a simultaneous increase in grasses and weeds and decrease in some kinds of trees owing to slash-and-burn forest clearing, field cropping, and herding. But at the same period blanket bogs creeping over the mountain slopes in oceanic parts of the British Isles were paludifying both pinelands and croplands (150). Britain ceased to be a sunny, wooded land well before the Romans came, in the Iron Age, both as invaders about fifty years B.C. and as settlers about a hundred years later.

Rising water tables associated with changes in climates and sea levels are documented by numerous paludification sequences in Europe and North America. In some parts of Britain, in the Late Bronze Age, brush and timber tracks were laid out over bog surfaces only to be covered over by bog growth, as were musical instruments buried in Ireland, Denmark, and Poland (and for some reason, ritualistic perhaps, intentionally broken before burial). Bog-covered artifacts of the Roman period, including datable coins, in Britain and Scandinavia, give evidence of a cool wetness that continued well into the Christian era (92, 100). In treeless peatlands of Denmark, scotch pine remains are buried deep in peat, and in Irish peatlands "bog oak" dug up from peat-buried forest remains is made into walking sticks and crucifixes. (Such "oak" remains also originated from trees other than *Quercus*.) Denmark's Tollund man (90) is well preserved in body and clothing (and even stomach contents). His bones are dissolved away, but his flesh and skin are tanned like leather. Remains of animals, some of them extinct, have been found in American prebog sediments, but few artifacts of human culture have showed up in bogs of glacial topography.

Paludification is linked to peatland formations across the boreal forest of North America, on old lake beds, including the plains once occupied by Glacial Lakes Agassiz, Upham, and Aiken in northern Minnesota and adjacent Manitoba, outwash plains, and interdunal depressions. The Lake Agassiz basin is now covered by 10–30 feet of peat. Forest peat began to build up there 4,000–5,000 years ago, and *Sphagnum* peat began to accumulate above the forest level 3,000 years ago. The continued development of bog vegetation up to the present time has been possible because the water table continued to rise as peat was deposited. Paludified depressions in Michigan's Upper Peninsula exhibit the usual fen-bog successional series, but jack pines are to be found in many of them, alone or in company with tamarack and black spruce.

Localized flooding may be linked to long-term climatic changes and also to beaver damming and destruction of forests by fire, disease, and clearing. A rapid development of peatlands may follow the loss of forest cover. The water draining through the soil can be increased as much as 40 percent as a consequence of forest removal because the soil receives more water and loses less (150). A forest canopy with a vast exposure of leaves intercepts and evaporates a tremendous quantity of water, and trees constantly pump water from the soil and lose it to the atmosphere by transpiration. Without tree cover a soil becomes wetter and less suited to renewed growth from tree seedlings. Forest regeneration is no longer possible, but a hydric succession of the sort that follows paludification is.

The lateral spread of peatlands from lake-fill mats has also been demonstrated (74). Paludification at the outer margin of bog mats can be attributed to downwash drainage blocked by impermeable peat.

Peatlands can also develop by the water-soaking of well-drained slopes. Blanket bogs can occupy a sloping terrain because of watertight soils such as fine-grained and compacted clay. Coarser soils are permeable at the surface, but downward percolation may bring about *podzolization*, a process by which oxides of iron and aluminum leach downward and accumulate as a hardpan. Pan formation is favored by an acid litter produced by conifers or heath plants. The accumulation of litter in itself causes water retention and oxygen-poor conditions that limit decomposition and favor bog initiation.

In the far north, permafrost reduces soil permeability and provides the waterlogging needed for peatland development.

The Peatland Climax, If Any

> In any weather, at any hour of the day or night, I have been anxious to improve the nick of time, and notch it on my stick too; to stand on the meeting of two eternities, the past and the future, which is precisely the present moment; to toe that line.
> —Thoreau

The black spruce stands at the outer margins of zoned peatlands can be considered muskeg, that is, a bog forest rooted in peat and having a single tree layer made up of *Picea mariana* and a hummocky ground topography covered by mosses, chiefly *Sphagnum*, as well as a shrub layer dominated by *Ledum*. This is the vegetation that characterizes the peatlands of the boreal forest. It is represented in the scattered peatlands of the Great Lakes region, though in a state of lesser development. (The term muskeg is sometimes used [204] more broadly to denote any wetland in which organic matter accumulates, in other words, any peatland regardless of vegetational cover or stage of development. In Canada and Alaska the term often denotes wetlands with more than one foot of accumulated peat.)

The muskeg develops in a zone of built-up peat offering a slight slope and a bit of drainage and aeration. The trees and the understory of *Ledum* alter the habitat by shading and increasing evapotranspiration. As *Sphagnum* growth is retarded and eventually checked, the peat is degraded. The high-bush associations of the lagg may then encroach on the muskeg to some limited extent because of a hydrology altered by peat loss. Decay in and near the lagg might conceivably allow some encroachment of white cedar from adjacent swampland. However, such an invasion would be possible only in localized areas under the influence of mineral soil and moving groundwater. Certainly *Thuja* is not suited to the edaphic and hydrological character of unaltered peat. It does not grow on bog peat low in oxygen and nutrients, although it sometimes survives the acid buildup of peat resulting from paludification.

Cedars grow in rich fens along watercourses and less commonly on lake mats consisting largely of calcareous lake sediments. At Summerby Swamp in Mackinac County, Michigan, an extensive rich fen and cedar swamp have built up in a highly calcareous area of some slope (as indicated by the rapid flow of Summerby Creek). Although *Sphagnum warnstorfii* and *S. fuscum* are common in large mounds, there is little continuous *Sphagnum* cover and scarcely any of the pioneer species, *Sphagnum* and otherwise, that initiate a fen-to-bog succession. Cedar swamps may also succeed rich fens marginal to shallow beach pools along the shores of Lake Michigan and Lake Huron where the water is rich in calcium and subject to fluctuating levels (figs. 2-23–25). A fringe of vegetation includes a number of calcicolous mosses disjunct from rich fens of the north (many of them found at Summerby Swamp as well). The mosses include *Scorpidium scorpioides, Drepanocladus intermedius, Campylium stellatum, Cinclidium stygium, Calliergon trifarium, Meesia uliginosa,* and *Catoscopium nigritum.* The liverworts *Preissia quad-*

rata, Moerckia hibernica, and *Aneura pinguis* are likewise calcicolous. In addition to a varied sedge growth, *Potentilla fruticosa, Pinguicula vulgaris, Selaginella selaginoides, Drosera rotundifolia,* and *D. linearis,* to mention but a few, give further evidence of mineral richness. The vegetation at the pool margins is quickly succeeded by a shrub growth of *Myrica, Salix,* and *Alnus* followed by *Larix* and *Thuja.* Mounds of *Sphagnum warnstorfii* and *S. fuscum* are found at the water's edge and at the interface between swamp and fen.

Nolten Lake, in Cheboygan County, Michigan (fig. 2-21), has a muddy false bottom exposed at the surface and covered with *Scirpus acutus* in the wettest parts and thin growths of *Carex lasiocarpa* and *C. oligosperma* in somewhat dryer zones. It is being invaded at the margins by *Larix* and *Thuja. Sphagnum teres, S. warnstorfii,* and *S. squarrosum* are sparsely represented there, together with such calciphiles as *Drepanocladus intermedius, Scorpidium scorpioides, Campylium stellatum, Tomenthypnum nitens, Cinclidium stygium,* and *Moerckia hibernica. Osmunda regalis, Phragmites australis,* and *Myrica gale* occupy the same fen-to-swamp transition. The shrubby Ericaceae have a scant presence in this kind of lake-fill.

At Ryerse Lake, in Mackinac County, is a large rich fen mat featuring some *Thuja* and some *Picea mariana. Sphagnum subsecundum* and *S. warnstorfii* in addition to *S. teres* are indicative of base-rich conditions, highly calcareous, in fact. The mat, like that at Nolten Lake, is developing into a cedar swamp.

Normally, at least, it seems that none of the wetlands in which *Thuja* occurs belong to a bog sequence, nor are they transitional to the forests of upland soils. *Thuja* never occurs in a bog with a previous moss-heath phase. Black spruce can grow in a cedar swamp, but *Thuja* cannot grow in a spruce muskeg. Seedlings of beech and sugar maple are sometimes seen in muskegs, but they never persist. Jack pines (fig. 2-26) grow on dry sands and also, rooted in sand, in paludified peatlands formed in old lake beds or dune slacks. White and red pines, sometimes very large, are occasional in bogs (fig. 2-27), anchored in mineral soil or on peat supported by a broad expanse of shallow roots. Both yellow and white birch grow in wet lowlands as well as dry uplands. Such examples of broad tolerance and freak occurrence do not demonstrate successional links between bogs, swamps, and upland forests. Moving, aerated groundwater and a high calcium content are essential for *Thuja*. Bogs have neither. Only a change in hydrological regime and a decomposition of acid peat can change a bog to a mineral-rich swamp. Only topographic change can transform a wetland to a well-drained upland. Low nighttime temperatures further deter upland species from invading lowlands and redirecting vegetational sequences.

Various authors (31, 48, 79) have theorized that bogs are linked by succession to an upland climax of beech, sugar maple, basswood, yellow birch, and hemlock. The hydrosere is seen as a change from open water to a floating fen mat community and then to a hummocky open bog dominated by leatherleaf and later black spruce, giving way eventually to white cedar. The drying of a cedar swamp as peat builds up is said to favor balsam fir and the eventual invasion by hardwoods. The transformation of a cedar swamp to an upland forest association is said to be very rare and dependent on profound changes in the environment. Of course, neither the wet nor the cold of lowland forests is likely to change, and therefore speculations on the unlikely seem profitless. It is true that some peat builds up in a cedar swamp, providing a substrate suitable for certain acidophiles, and a degree of floristic similarity exists between cedar swamps and the more nutrient-rich, early stages of bog development. It is also true that cedar swamps sometimes develop on "marly bog" lake mats. There is no evidence of acid bogs developing into cedar swamps.

Floristic comparisons of bogs and swamps have suffered from an uncertain definition of bogs. For example, Frank Gates's valuable accounts (75, 79, 82) of wetland communities and successions in northern Michigan made no attempt to separate *Thuja* swamps and fens from bogs. Even more recent literature on peatland chemistry has the same failing. Ordination, or ranking, of ninety "coniferous swamps" in northern Wisconsin (31) demonstrated a gradient from a pioneering *Carex* stage to a *Thuja-Abies* swamp stage, presumably by way of bog vegetation, but if the successions toward swamps and those toward bogs had been analyzed separately, would the acid phases have shown any similarity to the cedar swamps? The plants of a bog are not at all like those of a cedar swamp, although the lagg between them and a pioneering sedge mat may indeed share species with a cedar swamp. Choosing to analyze coniferous swamps in that area of Wisconsin may have weighted the results in favor of rich fen series. A comparison of the supposed spruce muskeg and cedar swamp linkage would be more meaningful than a successional study from fens of a diverse character. A study in northern lower Michigan has demonstrated very low similarity index values for coniferous swamps and bogs (223). It is hard to imagine a succession from wet, mineral-rich fens to relatively dry, mineral-poor muskegs followed by a reversion to wet and calcium-rich swamps.

Black spruce has been considered the climatic climax of the boreal forest (fig. 2-28) and the edaphic climax of peatlands farther south. It has also been regarded as a fire-regulated subclimax. But, to consider black spruce or any other forest type climactic in terrain so recently deglaciated is unreasonable. It is true that black spruce has its fullest development on the most mature peatlands and no other trees show signs of invading and replacing it. However, in the region of the upper Great Lakes, peatland successions are commonly incomplete and cyclic, and spruce muskegs are, at best, poorly developed. The habitat is marginal for black spruce and not at all suited to other trees. Black spruce trees make do with oxygen- and mineral-deficient soil, but they are stunted and slow growing. In Manitoba and Saskatchewan they may be as much as 100 feet tall, but in the bogs of the Great Lakes region they are much shorter, sometimes no more than 2 or 3 feet high. These unhealthy dwarfs are able to live in bogs because of shallow, spreading root systems, tolerance to environmental stress, and uncontested survival, not because of success in competition.

Peatland successions are as much cyclic as directional because of disturbance—fire, drought, inundation, wind-throw, and disease (figs. 2-29–30). By its layered growth, black spruce disturbs its own habitat. It alters the conditions suited to hummock growth and peat accumulation. It is at best short-lived. It is plagued by the parasitic dwarf mistletoe. It sinks into oozy peat where its roots are deprived of oxygen. But as it dies, it multiplies by layering, by sprouts formed on lower branches. And its offspring cause and endure a continued cycle of environmental changes.

In a more theoretical way, it is undesirable to think of a black spruce climax. In early stages of succession, photosynthesis exceeds respiration (that is, productivity exceeds decomposition), and organic detritus builds up. With changes in soil resulting from this imbalance in the energy budget, succession continues to a stable climax in which the energy produced equals the energy used. At this point, there is no further accumulation of undecomposed organic matter and no further succession. In a wet ecosystem where peat accumulates, however, anaerobic conditions continue to deter oxidative degradation, peat continues to accumulate, and the energy budget remains unbalanced. Accordingly, a black spruce muskeg, grounded on peat, is not a climax, no matter how stable a vegetation it may be.

Tamarack tolerates fire better than spruce, and the two sometimes grow intermingled in mats edaphically altered by burning. Stratigraphic studies give ample evidence of fires throughout the history of peatland development. Black spruce has a defense against fire in that some of its cones release seeds only after a burn. If the fire is severe, all the trees and their cones may be destroyed, and the chances of reestablishment are reduced by edaphic unsuitability and often by distance from a seed source. A lesser burn may destroy trees, yet favor seeding from fire-resistant cones. Actually, black spruce produces an abundance of seeds, but its seedlings are not abundant. (In much of the Great Lakes region, red squirrels drastically reduce the seed supply.) Seedling establishment requires a moist but unsaturated bed of living *Sphagnum* and also freedom from competition, but the proper conditions are not likely to follow burning. A fire that reduces the *Chamaedaphne* cover yet does not kill hummock-top *Sphagna* would favor seedling establishment. But *Chamaedaphne* readily survives fire, and if it does not, blueberries quickly take over and provide the same competition against spruce seedlings. The difficulties in reseeding may be a further reason why black spruce should be considered climax only by default, at least in the southern fringes of its range.

In summary, it can be said that black spruce indeed terminates succession on acid peat in the Upper Midwest, but it can scarcely be considered a climax. Cedar swamps terminate successions on a calcareous peat. Spruce and cedar belong to different kinds of hydroseres. They are not successional to one another or to upland forests of well-drained soils. The fumbly accounts of such successions seen in the literature are based on the imprecise use of wetland terminology—a failure to understand what a bog is and is not.

Fig. 2-1. Peatlands of the Hudson Bay Lowlands develop under cold desert conditions owing to impeded drainage and short, cool summers that allow precipitation to exceed evaporation and productivity to exceed decomposition. (Photo by Charles Tarnocai.)

Fig. 2-2. A boreal forest peatland dominated by black spruce muskeg, northeastern Manitoba. (Photo by Lynda Dredge.)

Fig. 2-3. An open peatland in northeastern Manitoba dominated by cotton grass, *Eriophorum*. (Photo by Lynda Dredge.)

Fig. 2-4. South Bog, Kalamazoo County, Michigan. The fen mat, nearly filling a lake basin about 20 feet in depth, is largely covered by a shrub-carr of *Aronia, Nemopanthus, Rhamnus, Vaccinium corymbosum,* and *Larix.* The ground layer of *Sphagnum* is topped by *Chamaedaphne.* (Photo by Janet Keogh and Richard Pippen.)

Fig. 2-5. Bog mat encroaching on a kettlehole lake in Somerset County, Maine. (Photo by Harry Tyler.)

Fig. 2-6. Lake-fill at Inverness Mud Lake, Cheboygan County, Michigan. The lake, originating as a bay of a more extensive postglacial lake, is nearly filled with muddy organic sediments. The sedge mat of *Carex lasiocarpa* is separated from a poor fen, or *Sphagnum* lawn, by a wooded "ice-push ridge" presumably caused by the pressure of ice floes blown by prevailing westerly winds against the floating mat. The open peatland at the right of the ice-push ridge consists of a *Sphagnum* lawn and toward the right a hummock-hollow complex covered by *Chamaedaphne* and in older phases spruce muskeg. (Photo by Gary Williams.)

Fig. 2-7. Peatland at Lake Sixteen, Cheboygan County, Michigan, formed as a result of lake-fill and marginal paludification

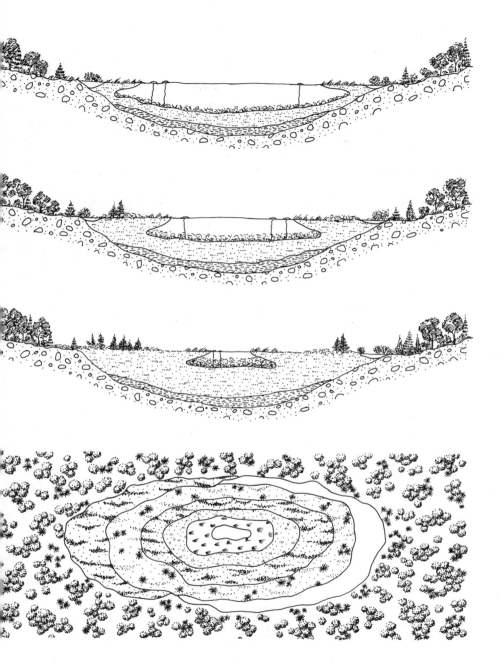

Fig. 2-8. Stages in the development of a lake-fill peatland showing the formation of a false bottom onto which a sedge mat encroaches as the pioneer stage in the succession from fen to bog. Because of blocked drainage as lake-fill progresses, new peatland may be formed external to the lagg owing to swamping, or paludification.

Fig. 2-9. A wet intermediate fen, relatively rich, at Smith's Fen, Cheboygan County, Michigan. The fen is dominated by *Carex lasiocarpa* and *Dulichium arundinaceum*. Burned in 1914 and again in 1917, this fen once supported a *C. lasiocarpa* mat encroaching on a pool and surrounded by a *Chamaedaphne-Sphagnum* complex and an outer stand of black spruce and white cedar. The fen is adjacent to a cedar swamp and apparently once merged with it.

Figs. 2-10–12. Pioneering stages of mat formation at an alkaline lake margin, Inverness Mud Lake, Cheboygan County, Michigan

Fig. 2-10.; False bottom sediments exposed in the *Nymphaea* zone in 1934. (Photo by Esther Rodgers Stevenson.)

Fig. 2-11. False bottom completely exposed, 1937. (Photo by Ralph Cooper.)

Fig. 2-12. *Carex* invading the exposed false bottom in 1977. (Photo by Gary Williams.)

Fig. 2-13. *Chamaedaphne* bog association at Inverness Mud Lake, Cheboygan County, Michigan. (Photo by Jerry Snider.)

Fig. 2-14. Clumps of black spruce formed by layering, Inverness Mud Lake. (Photo from the files of the University of Michigan Biological Station.)

Fig. 2-15. Black spruce muskeg, Luce County, Michigan. (Photo by Jerry Snider.)

Figs. 2-13–15. Stages in succession from the hummock-hollow, *Chamaedaphne*-dominated open bog to the black spruce muskeg developed on it

Fig. 2-16. Standing water forms a moat at the margin of the bog mat at Lake Sixteen, Cheboygan County, Michigan.

Fig. 2-17. *Typha latifolia*, cattail, indicative of marsh or fen habitats, is common in wet, open laggs, or moats. It is also commonly associated with moving water, at an inlet to a bog lake, for example, and it seems to be favored in sedge mats disturbed by foot traffic.

Fig. 2-18. Stutsmanville Bog, Emmet County, Michigan, an acid lake developed in a morainal depression, has a very narrow bog mat, with all stages of development telescoped together. An extensive black spruce muskeg in the background shows considerable mixing with tamarack near water's edge.

Fig. 2-19. Livingston Bog, Cheboygan County, Michigan, surrounding a small, acid lake of kettlehole origin and consisting of a narrow mat of *Chamaedaphne* and *Andromeda*. A clump of young black spruce trees is seen in the left foreground. (Photo from the files of the University of Michigan Biological Station.)

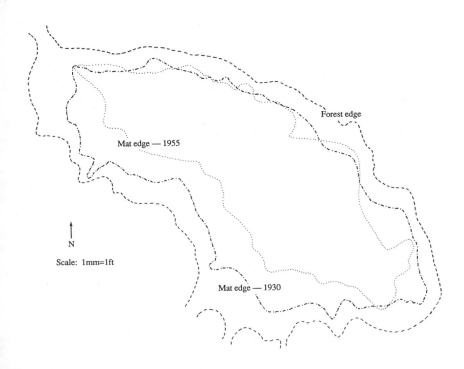

Fig. 2-20. A map showing the encroachment of the mat at Livingston Bog, Cheboygan County, Michigan, since 1930. Mapping was done by classes in plant ecology at the University of Michigan Biological Station.

Fig. 2-21. Nolten Fen, Cheboygan County, Michigan, with calcareous sediments exposed at the surface and supporting a thin growth of sedges and, at the outer margins, young *Thuja*. The vegetation is rather noticeably arranged in arcs. The open water is at the end of the lake receiving an inflow from underground, perhaps as seepage from Dingman Swamp.

Fig. 2-22. Nolten Fen, Cheboygan County, Michigan. The lake sediments exposed here are at least 15 feet deep. It is only at the surface that there is some structural support. This accumulation of mud is scarcely peat. Its very calcareous nature gives promise of a habitat suited to a cedar swamp already developing at the margins.

Fig. 2-23. In the early stages, brown mosses (such as *Drepanocladus*, *Campylium*, and *Scorpidium*) form a narrow fringe followed by shrubby willow, alder, and tamarack.

Figs. 2-23—25. Succession in and around shallow beach pools along Lake Huron at Evergreen Beach, Presque Isle County. (This is, incidentally, a fine habitat for rattlesnakes!)

Fig. 2-24. A marly sediment eventually exposed at the surface is invaded by sedges and rushes, finally *Thuja*.

Fig. 2-25. *Sphagnum warnstorfii* and *S. fuscum* form dense mounds at the transition from open fen to *Thuja* swamp.

Fig. 2-26. *Pinus banksiana*, jack pine, and *Chamaedaphne* in a paludified area near Trout Lake, Chippewa County, Michigan

Fig. 2-27. *Pinus strobus*, white pine, in sandy, paludified depression in Chippewa County, Michigan, occupied by peatland vegetation, including jack pine and leatherleaf

Fig. 2-28. A characteristic expression of *Picea mariana*, black spruce, with top-heavy branching, Hudson Bay Lowlands. Although black spruce is common in the Great Lakes region, it is dwarfed and short-lived there, in contrast to its development in the boreal forest. (Photo by Lynda Dredge.)

Fig. 2-29. Gates Bog, Cheboygan County, Michigan, in 1984. This bog, formed by paludification, was burned in 1918. It is now covered by a hummock-hollow, *Chamaedaphne* complex in its western two-thirds and a wet fen with a sparse growth of *Scirpus cyperinus* and sedges, as well as scattered clumps of *Chamaedaphne* in its more severely burned eastern third. There is no black spruce in the bog. Blueberries mingled with *Chamaedaphne* give evidence of fire. *Sphagnum cuspidatum* and *S. papillosum*, both indicative of relatively mineral-rich water, pioneer the wetness of the fen. Dense mats of *Polytrichum commune* are found at the junction of sandy slope and fen.

Fig. 2-30. Hoop Lake, Cheboygan County, Michigan. The lake is encroached on by a wet thicket of *Chamaedaphne* that shows disturbance and tree death because of paludification resulting from beaver activity.

Chapter 3

Peatland Plants and the Tonic of Wildness

> There too I admired, though I did not gather, the cranberries, small waxen gems, pendants of the meadow grass, pearly and red. . . .
> —Thoreau

Thoreau found the tonic of wildness and friendship of seasons while wading in marshes and smelling the whispering sedge. He spoke of marshes, meadows, and grass but admired in their midst the boggy cranberries and considered it a vulgar error to suppose one had tasted huckleberries who had never plucked them. It is too late to edit nomenclatural irregularities out of *Walden* and pointless to expect plain English to conform to scientific usage. Swamp, bog, and marsh are all one to Thoreau and to country folk, and so are grasses and sedges. A botanist may distinguish seedy huckleberries from blueberries, but to some of us a good huckleberry pie can be made of *Gaylussacia* or *Vaccinium*, or both. Cranberries do not grow in marshes, but Newfoundlanders call them marshberries. Nomenclature notwithstanding, the beauty of peatlands is there, a pleasure available to all but the insensitive and emotionally dry-shod. The plants of peatlands contribute much to the beauty and wonder that draw the rest of us ever downhill, toward mucky fens and bogs and wet feet.

Peatlands come in many kinds, and fens grade into bogs or swamps. Few species of peatlands are entirely specific in habitat requirements or tolerances. An acidophile on the Canadian Shield may well show a preference or tolerance for calcium in Minnesota and Michigan. But locally, at least, associations of species may serve better than individual species as habitat indicators. Certain species, such as *Menyanthes trifoliata* and *Sarracenia purpurea*, are sensitive indicators of even slight minerotrophy, even though they also occur in areas of considerable mineral enrichment. Some species, like *Carex sterilis*, *Juncus stygius*, *Tofieldia glutinosa*, and *Triglochin maritimum*, on the other hand, demand a high content in calcium and are accordingly limited to rich fens.

Rich Fens Transitional to Swamps

Species of rich fens in transition to *Thuja* swamps are extremely numerous, and many of them are good indicators of calcareous substrates, though not necessarily restricted to them. The following vascular plants are characteristic of rich fens in the upper Great Lakes region, at least:

Alnus rugosa
Aster junciformis
Calamagrostis canadensis
Calopogon tuberosus
Carex aquatilis
C. diandra
C. disperma
C. hystericina
C. lacustris
C. lanuginosa
C. lasiocarpa
C. leptalea
C. livida
C. sterilis
C. stricta
Cladium mariscoides
Cypripedium calceolus
Drosera linearis
D. rotundifolia
Eleocharis elliptica
E. pauciflora
E. rostellata
Epilobium palustre
Eriocaulon septangulare
Eriophorum viridi-carinatum
Gentiana procera
Glyceria striata
Juncus balticus
J. stygius
Juniperus horizontalis
Larix laricina
Liparis loeselii
Lobelia kalmii
Lonicera oblongifolia
Lycopodium inundatum
Muhlenbergia glomerata
Myrica gale
Parnassia glauca
Phragmites australis
Pinguicula vulgaris
Potentilla fruticosa
Pyrola asarifolia
Rhamnus alnifolia
Rhynchospora fusca
Salix candida
Sarracenia purpurea
Scirpus acutus
S. cespitosus
S. hudsonianus
Selaginella selaginoides
Senecio pauperculus
Solidago uliginosa
Tofieldia glutinosa
Triglochin maritimum
Valeriana uliginosa
Zigadenus glaucus

Sphagnum warnstorfii is exceedingly common in rich fens, and *S. fuscum* forms large mounds somewhat above the level of *S. warnstorfii*. *Picea mariana*, so characteristic of acid bogs, manages well on peaty mounds of *Sphagnum*. The calcareous wetness favors numerous "brown mosses"—*Tomenthypnum nitens, Calliergon trifarium, Campylium stellatum, Drepanocladus intermedius, Scorpidium scorpioides, Cinclidium stygium,* and *Catoscopium nigritum.* Aneura pinguis and *Moerckia hibernica,* both thallose liverworts, are equally indicative of calcium enrichment.

Typha latifolia is common in intermediate and rich fens and also in wet laggs (figs. 2-16–17). It is an indicator of mineral-rich, aerated habitats, especially where there is movement of groundwater.

Empetrum nigrum, the black crowberry (fig. 3-66), is found as a rarity in a marly rich fen just north of the Straits of Mackinac, in northern Michigan. In its more characteristic range, far north of the Great Lakes region, it is not restricted to such calcium-rich areas.

In *Thuja* swamps is a rich flora at the field layer, with many mosses and herbs or low-growing woody species, most notably, *Mitella nuda, Trientalis borealis, Galium triflorum, Coptis trifolia, Linnaea borealis, Clintonia borealis, Rubus pubescens, Gaultheria hispidula, Carex gynocrates, C. leptalea, C. interior, C. trisperma, Glyceria striata, Lycopus uniflorus, Pyrola secunda, Prunella vulgaris, Aralia nudicaulis, Epilobium glandulosum,* and *Equisetum scirpoides.*

The Fen-to-Bog Sequence

The sedge fens antecedent to bogs range from an intermediately rich to poor water chemistry. At the margin of alkaline lakes the pioneering species of intermediate fens are indicative of mineral-rich conditions. In poor fens bordering acid lakes the plants have a lesser need or tolerance for minerals in solution. As fens become open bogs and spruce muskegs, the kinds of plants as well as their frequencies and associations shift in response to changes in moisture, temperature, aeration, nutrient supply, pH, and exposure. No two fens, no two bogs are alike, and the zones that link them in a developmental sequence lack definable boundaries.

The "Alkaline Edge" Mat

The pioneering mat pushing onto a lake of alkalinity, or at least mineral richness, is dominated by sedges. It is an *intermediate fen,* well nourished, well aerated, rich in species. Productivity far exceeds decay, and peat accumulation alters the mat in wetness and mineral richness. Those alterations are supplemented by the acidity brought on by *Sphagnum* invaders. The sedge mat is influenced by the lake water and its content in calcium and bicarbonate ions. The pH of the lake water, however rich in nutrients, is buffered at 7 to 8. The water is made murky by a superabundance of planktonic algae and invertebrates, dead and alive. A muddy suspension of organic matter creates a false bottom on which floating leaved macrophytes like *Nuphar*, *Nymphaea* (fig. 3-4), and *Potamogeton* come to be rooted and onto which sedges advance as drought exposes the muddiness to drying and compacting.

The sedge mat is wet. It floats. It quakes. *Carex lasiocarpa* is the dominant mat former. Its interwoven rhizomes provide a firm though buoyant support for other pioneers. *Carex limosa* and *C. chordorrhiza* also form mats by sprouting at the nodes of culms fallen into the water. *Cladium mariscoides* may grow with *Carex* or separate from it in similar, though perhaps richer, situations. The mat is typically open and sedgy, but *Chamaedaphne* and *Andromeda* can provide shrub cover up to the water's edge. *Alnus rugosa* often flourishes on the mat, near the mouth of inlets that may be bordered or even choked by *Typha latifolia*. Certain wet lagg shrubs—*Aronia*, *Ilex*, *Rhamnus*, *Myrica*, and *Salix*—commonly grow with *Alnus*. *Potentilla palustris*, *Vaccinium macrocarpon*, and *Menyanthes trifoliata* are common. Other highly characteristic species include:

Calamagrostis canadensis	*Lysimachia terrestris*
Calla palustris	*Muhlenbergia glomerata*
Campanula aparinoides	*Myrica gale*
Carex aquatilis	*Onoclea sensibilis*
C. cephalantha	*Osmunda regalis*
C. chordorrhiza	*Phragmites australis*
C. comosa	*Pogonia ophioglossoides*
C. limosa	*Sarracenia purpurea*
C. stricta	*Thelypteris palustris*
Drosera rotundifolia	*Triadenum fraseri*
Geum rivale	*Viola cucullata*
Iris versicolor	*V. nephrophylla*

These and other species of more incidental occurrence contribute to a diversity that falls off abruptly in each of the subsequent zones of vegetation. (A number of these plants grow equally well in other habitats of mineral enrichment, such as laggs and rich fens.)

Sphagnum teres is the peat moss of prime significance in the sedge mat. (*Sphagnum subsecundum* and, in north-central Minnesota and presumably northward, *S. subsecundum* var. *contortum* and *S. obtusum* also pioneer rich fen-to-swamp successions.) A few mosses, most notably *Calliergonella cuspidata* and *Calliergon stramineum*, hide away at the wet bases of sedges. The thallose liverwort *Moerckia hibernica* may grow there too.

The Sphagnum Lawn Community

Back of the sedge mat is an open *poor fen*, or *Sphagnum* lawn, also wet and floating. It is relatively acid, but its nutrient status is enhanced by contact with groundwater. It lacks topographic relief except near the hinge line between floating mat and grounded bog where hummock formation favors a scattered growth of *Larix*, often in association with *Betula pumila*. The carpet is dominated by a few species of *Sphagnum*, *S. cuspidatum* (and sometimes *S. majus*) in soaks and puddles, *S. papillosum* barely above standing water, and, slightly higher than that, the hummock-forming *S. magellanicum* and *S. capillifolium*. *Chamaedaphne* can be present. *Andromeda* is best represented in the wetness of this zone.

The *Sphagnum* lawn is not conspicuously sedgy, but *Rhynchospora alba*, *Carex oligosperma*, *C. limosa*, *C. pauciflora*, *Eriophorum virginicum*, and *E. angustifolium* may be present, together with the cranberries, *Vaccinium macrocarpon*, and, more characteristically, *V. oxycoccos*. *Scheuchzeria* is abundant in especially wet sites, and *Utricularia cornuta* sometimes makes a show in puddles. The orchids *Calopogon tuberosus*, *Habenaria clavellata* (and less commonly *H. blephariglottis*), *Pogonia ophioglossoides*, and *Arethusa bulbosa* occupy this zone, as do the carnivores *Sarracenia purpurea*, *Drosera intermedia*, and *D. rotundifolia*. In upper Michigan, at least, *Carex exilis* is rare, and *Xyris montana* is represented in a spotty way. Holdovers from the sedge mat phase of succession, such as *Menyanthes*, add diversity.

The leafy liverwort *Cladopodiella fluitans* grows in soaks, in soppy black masses. The moss *Drepanocladus fluitans* may occur in depressions subject to drying, and *Polytrichum strictum* is often abundant, though scattered.

The Grounded, Sphagnum Hummock, Low-Shrub Mat

The bog is grounded on peat. Highly acid and poor in nutrients, it consists of a low-shrub, hummock-hollow zone and an older black spruce muskeg. The low-shrub zone is dominated at ground level by *Sphagnum recurvum* in dryer hollows, *S. capillifolium*, *S. magellanicum*, and *S. fuscum* in hummocks. Above the hummocks is a layer of *Chamaedaphne*, a long-lived, clonal shrub continually forming adventitious roots and proliferating sprouts above the level of saturation. It is not easily destroyed by fire, but in bogs burned over blueberries (*Vaccinium myrtilloides* and *V. angustifolium*) may crowd in. *Kalmia polifolia* belongs in this zone. *Eriophorum virginicum*, *E. spissum*, *E. angustifolium*, and *Carex oligosperma* may be fairly common, *C. limosa* and *C. paupercula* less so. Tamaracks often linger on as relics from the *Sphagnum* lawn, often in disturbed areas in regression from bog to fen, and black spruce occurs in scattered clumps. The spruce trees are commonly parasitized by the dwarf mistletoe *Arceuthobium*. *Carex trisperma* favors the shade of spruce and tamarack, usually occurring on drier mounds.

The *Sphagnum* hummocks are often covered at maturity by a few lichens, species of *Cladina* and *Cladonia*, and the true mosses, *Polytrichum strictum*, *Dicranum undulatum*, *Pohlia nutans*, and *Pleurozium schreberi*. The hollows are relatively dry, except under spruce clumps where several leafy liverworts—*Mylia anomala*, *Kurtzia setacea*, *Calypogeia sphagnicola*, and *Cladopodiella fluitans*—have a furtive existence in dark, wet holes.

The Black Spruce Muskeg

The spruce muskeg is poorly represented in the Great Lakes states and, because of a vegetational continuum with the low-shrub zone, often difficult to characterize. It is relatively dry, highly acid, poor in nutrients and species, and shady. *Chamaedaphne* and *Kalmia* are largely replaced in the shade by *Ledum* and *Vaccinium angustifolium*. *Sphagnum magellanicum* and *S. recurvum* provide a nearly continuous ground cover, and some true mosses, *Dicranum polysetum* and the feather mosses *Pleurozium schreberi*, *Hylocomium splendens*, and *Ptilium crista-castrensis*, have some prominence. Vascular plants at ground level include *Carex trisperma*, *Clintonia borealis*, *Coptis trifolia*, *Cypripedium acaule*, *Gaultheria hispidula*, *Maianthemum canadense*, *Monotropa uniflora*, *Rubus pubescens*, and *Smilacina trifolia*.

The Lagg

At the outer margin, drainage from the bog convexity, however slight, and the surrounding slopes may create a wet lagg of mineral richness. The lagg may be open and occupied by shallow water (as a moat), a sedge fen, or a dense fern growth of *Osmunda regalis* or *O. cinnamomea*, or it may be closed in by high bushes, *Nemopanthus mucronatus*, *Aronia prunifolia*, *Rhamnus alnifolia*, *Ilex verticillata*, and *Viburnum cassinoides*. A sedgy lagg may have a landward margin of *Myrica* or *Salix*. The shrubby willows including *Salix candida*, *S. petiolaris*, *S. pedicellaris*, *S. pyrifolia*, and *S. serissima*, may occur here and also at the rich-fen margins of beach pools and in other marly habitats. *Alnus* often dominates the wet lagg. Herbs of the wet lagg include *Calla palustris*, *Carex trisperma*, *C. rostrata*, *C. lasiocarpa*, and *C. stricta* in association with *Glyceria striata*, *G. canadensis*, *Iris versicolor*, *Lycopus uniflorus*, *Scirpus cyperinus*, and *Thelypteris palustris*. *Sphagnum teres* and *S. subsecundum* are sometimes present, especially in alder-dominated laggs. Downed logs from adjacent woodlands contribute habitats favorable to mosses, and *Aulacomnium palustre* and *Hypnum lindbergii* give evidence of mineral-rich wetness at ground level. *Polytrichum commune* often grows where cleared slopes join an open wet lagg.

The lagg is often poorly marked in the Great Lakes states, because the bogs are rarely raised enough to provide much runoff drainage, because they are commonly bordered by swampy lowlands rather than wooded slopes, or because drainage from sterile upland soils is mineral poor. In a relatively dry lagg the muskeg sometimes merges with a shrub growth of *Gaylussacia baccata, Vaccinium angustifolium, V. myrtilloides,* and (southward) *V. corymbosum.* Next to a cedar swamp a wet alder community provides transition to woodland. The swamp-growing *Sphagnum centrale, S. warnstorfii, S. teres, S. squarrosum, S. russowii, S. girgensohnii,* and *S. fimbriatum* grow there in abundance, as do many true mosses of wet habitats, *Calliergon cordifolium* and *C. giganteum,* for example. In this zone, as much swamp as lagg, are numerous other plants characteristic of *Thuja* swamps, even *Thuja* itself.

The Acid Edge Sequence

A *Sphagnum* lawn develops at the edge of acid, mineral-poor lakes. Such lakes have the relatively clear water and firm bottoms associated with low productivity. Aquatic macrophytes are less conspicuous than in alkaline lakes. The narrow mat is often pioneered by *Sphagnum cuspidatum* growing out into the water and also forming wet masses just back of the edge. It is invaded by *S. papillosum,* which quickly gives way to *S. magellanicum* and *S. capillifolium.* (*S. jensenii* and *S. majus* may also serve as aquatic pioneers.) Sedges grow sparsely among the *Sphagna* and sometimes form thin stands in shallow water beyond the mat, as *Andromeda* and *Chamaedaphne* also do. *Rhynchospora alba* and *Carex limosa* are common. The mat abruptly gives way to a zone where *Ledum* and *Larix* clump with *Picea mariana* in a narrow transition to a parklike muskeg. The species diversity is reduced as compared to that of a *Sphagnum* lawn successional to a mineral-rich sedge mat.

Paludified Lowlands

In peatlands developed in paludified lowlands rather than lake basins, the fen stages may be relatively rich and sedgy or poor with a greater abundance of *Sphagnum* and other acidophiles. The richer fens often have a good show of *Sphagnum subsecundum* and *S. pulchrum,* and the poorer ones have an abundance of *S. majus, S. papillosum, S. recurvum,* and *S. riparium.* Hummock formation involves the same red species, *S. magellanicum* and *S. capillifolium,* and the same brown cap of *S. fuscum* as in other peatlands.

Characteristic Vascular Plants of Peatlands, Including Rich Fens in Transition to *Thuja* Swamps

Pteridophytes
  Osmundaceae
    Osmunda cinnamomea L. (fig. 3-27)   Cinnamon fern
    O. regalis L. (fig. 3-26)   Royal fern
  Aspleniaceae
    Onoclea sensibilis L. (fig. 3-29)   Sensitive fern
    Thelypteris palustris Schott (fig. 3-28)   Marsh fern
  Selaginellaceae
    Selaginella selaginoides Link   Northern spikemoss
  Lycopodiaceae
    Lycopodium inundatum L.   Clubmoss

Gymnosperms
  Pinaceae
    Abies balsamea (L.) Miller (fig. 3-35)   Balsam fir
    Larix laricina (DuRoi) K. Koch (fig. 3-32)   Tamarack
    Picea glauca (Moench) A. Voss (fig. 3-31)   White spruce
    P. mariana (Miller) BSP (fig. 3-30)   Black spruce
    Pinus banksiana Lamb. (fig. 3-33)   Jack pine
  Cupressaceae
    Juniperus horizontalis Moench   Creeping juniper
    Thuja occidentalis L. (fig. 3-34)   White cedar

Angiosperms
  Typhaceae
    Typha latifolia L. (fig. 2-17)   Common cattail
  Sparganiaceae
    Sparganium minimum (Hartm.) Fries   Bur reed
  Juncaginaceae
    Scheuchzeria palustris L. (figs. 3-8, 3-113)
    Triglochin maritimum L. (fig. 3-111)   Arrow grass
    T. palustre L. (fig. 3-112)   Arrow grass
  Gramineae
    Calamagrostis canadensis (Michaux) Beauv. (fig. 3-107)   Bluejoint
    Glyceria canadensis (Michaux) Trin. (fig. 3-109)   Rattlesnake grass
    G. striata (Lam.) Hitchc. (figs. 3-3, 3-110)   Fowl manna grass
    Muhlenbergia glomerata (Willd.) Trin. (fig. 3-108)   Marsh wild timothy
    Phragmites australis (Cav.) Steudel (figs. 3-13, 3-106)   Reed

Cyperaceae
  Carex aquatilis Wahl.   Species of *Carex* are
    (fig. 3-68)   commonly known as
  C. canescens L. (fig. 3-75)   sedges.
  C. capillaris L. (fig. 3-83)
  C. cephalantha (Bailey) Bickn.
    (fig. 3-89)
  C. chordorrhiza L. f. (fig. 3-85)
  C. comosa Boott (fig. 3-69)
  C. diandra Schrank (fig. 3-76)
  C. disperma Dewey (fig. 3-78)
  C. exilis Dewey (fig. 3-90)
  C. flava L. (fig. 3-82)
  C. gynocrates Drejer (fig. 3-86)
  C. hystericina Willd. (fig. 3-7)
  C. interior Bailey (fig. 3-88)
  C. lacustris Willd. (fig. 3-67)
  C. lanuginosa Michaux
  C. lasiocarpa Ehrh. (fig. 3-79)
  C. leptalea Wahl. (fig. 3-84)
  C. limosa L. (fig. 3-81)
  C. livida (Wahl.) Willd.
    (fig. 3-74)
  C. oligocarpa Willd.
  C. oligosperma Michaux
    (fig. 3-71)
  C. pauciflora Lightf. (fig. 3-70)
  C. paupercula Michaux
    (fig. 3-80)
  C. rostrata Stokes (fig. 3-73)
  C. sterilis Willd. (fig. 3-87)
  C. stricta Lam. (fig. 3-72)
  C. trisperma Dewey (fig. 3-77)
  Cladium mariscoides (Muhl.) Torrey   Twig rush
    (fig. 3-102)
  Dulichium arundinaceum (L.) Britton   Three-way sedge
    (figs. 3-9–10, 3-103)
  Eleocharis elliptica Kunth (fig. 3-105)   Spike rush
  E. pauciflora (Lightf.) Link   Spike rush
  E. rostellata Torrey   Spike rush
  Eriophorum angustifolium Honck.   Tall cotton grass
    (fig. 3-98)

E. spissum Fern. (fig. 3-95) — Hare's tail
E. virginicum L. (figs. 3-2, 3-96) — Tawny cotton grass
E. viridi-carinatum (Engelm.) Fern. (figs. 3-1, 3-97) — Cotton grass
Rhynchospora alba (L.) Vahl (fig. 3-94) — White beak rush
R. fusca (L.) Ait. f. — Brown beak rush
Scirpus acutus Bigelow (fig. 3-91) — Hardstem bulrush
S. cespitosus L. (fig. 3-100) — Deer grass
S. cyperinus (L.) Kunth (fig. 3-93) — Wool grass
S. hudsonianus (Michaux) Fern. (figs. 3-6, 3-99) — Alpine cotton grass
S. validus Vahl (fig. 3-92) — Softstem bulrush

Araceae
Calla palustris L. (figs. 3-5, 3-126) — Wild calla

Xyridaceae
Xyris difformis Chapman — Yellow-eyed grass
X. montana Ries (fig. 3-101) — Yellow-eyed grass

Eriocaulaceae
Eriocaulon septangulare With. (fig. 3-11) — Pipewort

Juncaceae
Juncus balticus Willd. (fig. 3-104) — Rush
J. stygius L. — Rush

Liliaceae
Clintonia borealis (Ait.) Raf. — Bluebead lily
Smilacina trifolia (L.) Desf. (fig. 3-125) — False Solomon's seal
Tofieldia glutinosa (Michaux) Pers. (fig. 3-114) — False asphodel
Zigadenus glaucus (Nutt.) Nutt. — White camas

Iridaceae
Iris versicolor L. (fig. 3-128) — Wild blue flag

Orchidaceae
Arethusa bulbosa L. (fig. 3-120) — Dragon's mouth
Calopogon tuberosus (L.) BSP. (fig. 3-122) — Grass pink
Cypripedium acaule Ait. (fig. 3-123) — Stemless lady slipper
C. calceolus L. (fig. 3-124) — Yellow lady slipper
Habenaria blephariglottis (Willd.) Hooker (fig. 3-117) — White fringed orchid
H. clavellata (Michaux) Sprengel (fig. 3-119) — Club-spur orchid
Liparis loeselii (L.) Richard (fig. 3-118) — Fen orchid
Pogonia ophioglossoides (L.) Ker (fig. 3-121) — Rose pogonia

Salicaceae
Salix candida Willd. (fig. 3-37) — Hoary willow

S. pedicellaris Pursh (fig. 3-38) — Bog willow
S. petiolaris J. E. Smith (fig. 3-39) — Meadow willow
S. pyrifolia Andersson (fig. 3-41) — Balsam willow
S. serissima (Bailey) Fern. (fig. 3-40) — Autumn willow

Myricaceae
Myrica gale L. (figs. 3-19–20, 3-36) — Sweet gale

Betulaceae
Alnus rugosa (Duroi) Sprengel (fig. 1-11) — Speckled alder
B. pumila L. (fig. 1-13) — Bog birch

Viscaceae
Arceuthobium pusillum C. H. Peck (figs. 4-13–15) — Dwarf mistletoe

Ranunculaceae
Coptis trifolia (L.) Salisb. — Goldthread

Droseraceae
Drosera intermedia Hayne (figs. 4-5, 4-7) — Sundew
D. linearis Goldie (fig. 4-4) — Sundew
D. rotundifolia L. (fig. 4-3) — Round-leaved sundew

Sarraceniaceae
Sarracenia purpurea L. (figs. 4-2, 4-10) — Pitcher plant

Saxifragaceae
Parnassia glauca Raf. (fig. 3-115) — Grass of Parnassus

Rosaceae
Aronia prunifolia (Marsh.) Rehder (figs. 3-21, 3-54) — Chokeberry
Geum rivale L. — Avens
Potentilla fruticosa L. (fig. 3-45) — Shrubby cinquefoil
P. palustris (L.) Scop. (fig. 3-48) — Marsh cinquefoil
Rubus pubescens Raf. (fig. 3-15) — Dwarf raspberry
Spiraea alba Duroi (figs. 3-12, 3-46) — Meadowsweet

Empetraceae
Empetrum nigrum L. (fig. 3-66) — Black crowberry

Anacardiaceae
Toxicodendron vernix (L.) Kuntze (fig. 3-44) — Poison sumac

Aquifoliaceae
Ilex verticillata (L.) A. Gray (fig. 3-50) — Michigan holly
Nemopanthus mucronatus (L.) Loes. (figs. 3-17, 3-49) — Mountain holly

Aceraceae
Acer rubrum L. (fig. 1-12) — Red maple

Rhamnaceae
Rhamnus alnifolia L'Hér. (fig. 3-43) — Alder-leaved buckthorn

Guttiferae
   Triadenum fraseri (Spach) Gl.   Marsh St. Johnswort
Violaceae
   Viola cucullata Ait.   Marsh violet
   V. nephrophylla Greene   Violet
Onagraceae
   Epilobium palustre L.
Cornaceae
   Cornus stolonifera Michaux (fig. 3-53)   Red osier
Pyrolaceae
   Pyrola asarifolia Michaux (fig. 3-116)   Shinleaf
   P. secunda L.   Shinleaf
Ericaceae
   Andromeda glaucophylla Link. (figs. 3-23, 3-57)   Bog rosemary
   Chamaedaphne calyculata (L.) Moench. (figs. 3-25, 3-58)   Leatherleaf
   Gaultheria hispidula (L.) Muhl. (figs. 3-14, 3-65)   Creeping snowberry
   Gaylussacia baccata (Wang.) K. Koch (fig. 3-62)   Huckleberry
   Kalmia angustifolia L.   Sheep laurel
   K. polifolia Wang. (figs. 3-24, 3-56)   Swamp laurel
   Ledum groenlandicum Oeder (figs. 3-22, 3-55)   Labrador tea
   Vaccinium angustifolium Ait. (fig. 3-61)   Blueberry
   V. macrocarpon Ait. (fig. 3-63)   Cranberry
   V. corymbosum L. (fig. 3-59)   Highbush blueberry
   V. myrtilloides Ait. (figs. 3-16, 3-60)   Velvet-leaf blueberry
   V. oxycoccos L. (fig. 3-64)   Small cranberry
Primulaceae
   Lysimachia terrestris (L.) BSP.   Yellow loosestrife
   L. thyrsiflora Gray   Yellow loosestrife
Gentianaceae
   Gentiana procera Holm.   Gentian
   Menyanthes trifoliata L. (fig. 3-127)   Buckbean
Labiatae
   Lycopus uniflorus Michaux   Bugleweed
Lentibulariaceae
   Pinguicula vulgaris L. (figs. 4-6, 4-11)   Butterwort
   Utricularia cornuta Michaux   Bladderwort
   U. intermedia Hayne (fig. 4-9)   Bladderwort

Caprifoliaceae
   Lonicera oblongifolia (Goldie) Hooker   Fly honeysuckle
     (fig. 3-51)
   L. villosa (Michaux) R. & S.   Fly honeysuckle
   Viburnum cassinoides L. (figs. 3-18, 3-42)   Withe rod
Valerianaceae
   Valeriana sitchensis subsp. *uliginosa*   Valerian
     (Torrey & Gray) F. G. Meyer
Campanulaceae
   Campanula aparinoides L.   Marsh bellflower
Lobeliaceae
   Lobelia kalmii L.   Lobelia
Compositae
   Aster junciformis Rydberg   Aster
   Senecio pauperculus Michaux   Groundsel
   Solidago uliginosa Nutt.   Goldenrod

Fig. 3-1. *Eriophorum viridi-carinatum*, tall cotton grass, is a rich fen species.

Fig. 3-2. *Eriophorum virginicum*, tawny cotton grass, common in poor fens. Wide-spreading bracts subtend the inflorescence.

Fig. 3-3. *Glyceria striata*, fowl manna grass, is indicative of mineral content in the soil water.

Fig. 3-4. *Nymphaea odorata*, water lily, an aquatic macrophyte that thrives in mineral-rich waters, has white flowers. (Photo by Gary Williams.)

Fig. 3-5. *Calla palustris*, wild calla, grows in mineral-rich laggs.

Fig. 3-6. *Scirpus hudsonianus*, alpine cotton grass, of rich fens, has a cottony inflorescence suggestive of *Eriophorum*.

Fig. 3-7. *Carex hystericina*, a calciphilic sedge, is common in rich fens and cedar swamps.

Fig. 3-8. *Scheuchzeria palustris* belongs in especially wet, poor fen habitats.

Fig. 3-9. *Dulichium arundinaceum*, the 3-way sedge, in an intermediate fen habitat

Fig. 3-10. *Dulichium arundinaceum*, viewed from above

Fig. 3-11. *Eriocaulon septangulare*, rooted on calcareous mud. Several species of *Eriocaulon* are among the first invaders of false-bottom exposures.

Fig. 3-12. *Spiraea alba,* meadowsweet

Fig. 3-13. *Phragmites australis,* reed. The leaf blades are broad and wide-spreading pennants.

Fig. 3-14. *Gaultheria hispidula*, creeping snowberry, on a bed of *Sphagnum magellanicum* in a spruce muskeg

Fig. 3-15. *Rubus pubescens*, dwarf raspberry

Fig. 3-16. *Vaccinium myrtilloides*, blueberry. The leaves are rugose, and both leaves and branchlets are fuzzy.

Fig. 3-17. *Nemopanthus mucronatus*, mountain holly, has leaves ending in a minute bristle.

Fig. 3-18. *Viburnum cassinoides*, withe rod, is made striking by multicolored berries.

Fig. 3-19

Fig. 3-20

Fig. 3-21. *Aronia prunifolia*, chokeberry, has beaded-serrate leaf margins.

Figs. 3-19–20. *Myrica gale*, sweet gale, once used in England for delousing beds, has nitrogen-fixing root nodules.

Fig. 3-22. *Ledum groenlandicum,* Labrador tea

Fig. 3-23. *Andromeda glaucophylla,* bog rosemary

Fig. 3-24. *Kalmia polifolia,* swamp laurel

Fig. 3-25. *Chamaedaphne calyculata,* leatherleaf, in fruit

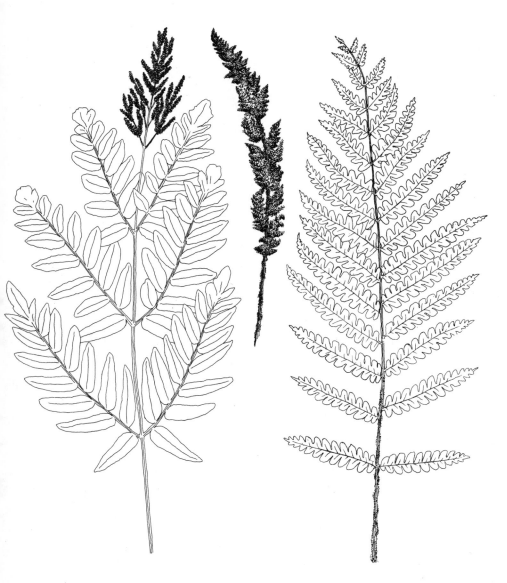

Fig. 3-26. *Osmunda regalis*, royal fern

Fig. 3-27. *O. cinnamomea*, cinnamon fern, with spore-bearing leaf separate

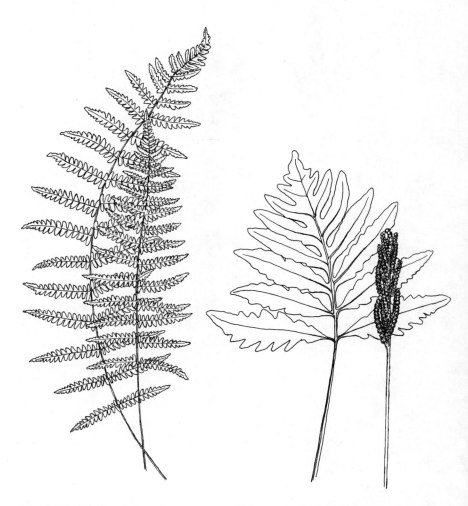

Fig. 3-28. *Thelypteris palustris,* marsh shield fern

Fig. 3-29. *Onoclea sensibilis,* sensitive fern, so named because the fronds are sensitive to early frost

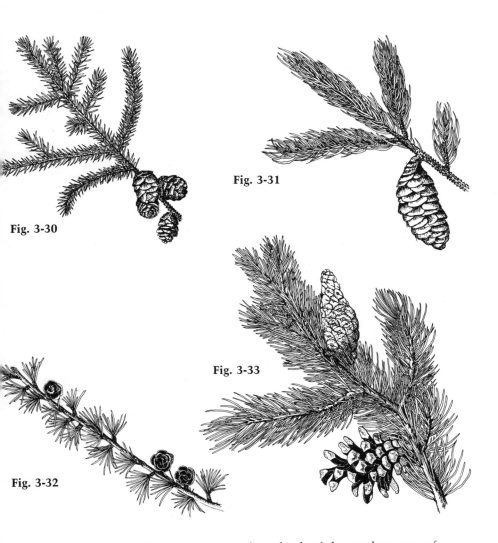

Figs. 3-30–33. The conifers of swamps and peatlands of the northern part of the Great Lakes region. **Fig. 3-30.** *Picea mariana*, black spruce. **Fig. 3-31.** *Picea glauca*, white spruce. **Fig. 3-32.** *Larix laricina*, tamarack. **Fig. 3-33.** *Pinus banksiana*, jack pine.

Fig. 3-34. *Thuja occidentalis*, northern white cedar

Fig. 3-35. *Abies balsamea*, balsam fir

Fig. 3-36. *Myrica gale*, sweet gale, is a low-growing shrub often found at the landward margins of bog mats and in open rich fens.

Fig. 3-37. *Salix candida*, hoary willow, has leaves that are white underneath. It grows in a variety of mineral-rich habitats, such as the margins of open fens and marshes, as do other willows.

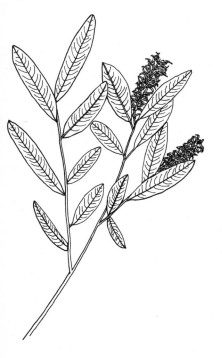

Fig. 3-38. *Salix pedicellaris*, bog willow

Fig. 3-39. *Salix petiolaris*, meadow willow

Fig. 3-40. *Salix serissima*, autumn willow

Fig. 3-41. *S. pyrifolia*, balsam willow

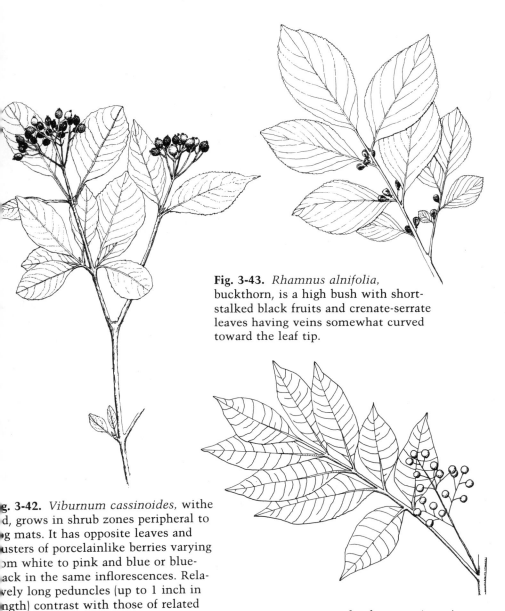

Fig. 3-43. *Rhamnus alnifolia*, buckthorn, is a high bush with short-stalked black fruits and crenate-serrate leaves having veins somewhat curved toward the leaf tip.

g. 3-42. *Viburnum cassinoides*, withe d, grows in shrub zones peripheral to g mats. It has opposite leaves and usters of porcelainlike berries varying om white to pink and blue or blue-ack in the same inflorescences. Rela-vely long peduncles (up to 1 inch in ngth) contrast with those of related ecies.

Fig. 3-44. *Toxicodendron vernix*, poison sumac, is a rather tall shrub constantly associated with tamarack in the southern part of the Great Lakes region. Autumn-red leaves and waxy-white berries aid in recognition.

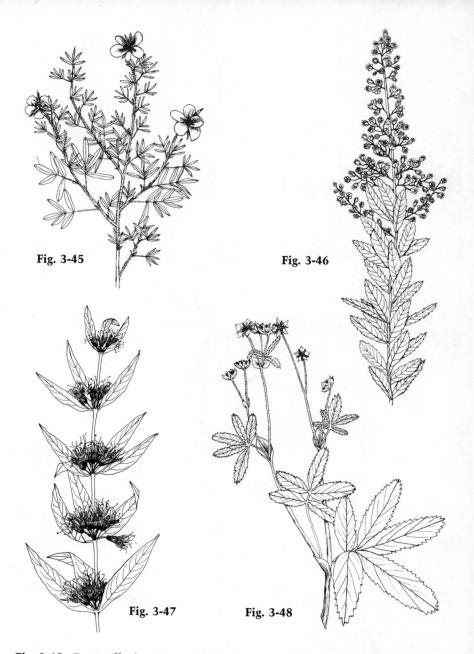

Fig. 3-45. *Potentilla fruticosa*, shrubby cinquefoil, a low bush with shreddy-brown bark, pubescent leaves, and bright yellow flowers, grows in rich fens. **Fig. 3-46.** *Spiraea alba*, meadowsweet, with spires of small, white flowers, grows as a low shrub at margins of sedgy fens. **Fig. 3-47.** *Decodon verticillatus*, water willow or swamp loosestrife, is a shrubby perennial that dies back each fall. It has whorled leaves with showy magenta flowers in their axils. The plant grows at the margins of ponds and in shrubby swamps, especially in the southern part of the Great Lakes region. **Fig. 3-48.** *Potentilla palustris*, marsh cinquefoil, has dark maroon flowers. It is a low shrub, less than knee high, growing in wet, mineral-rich sites.

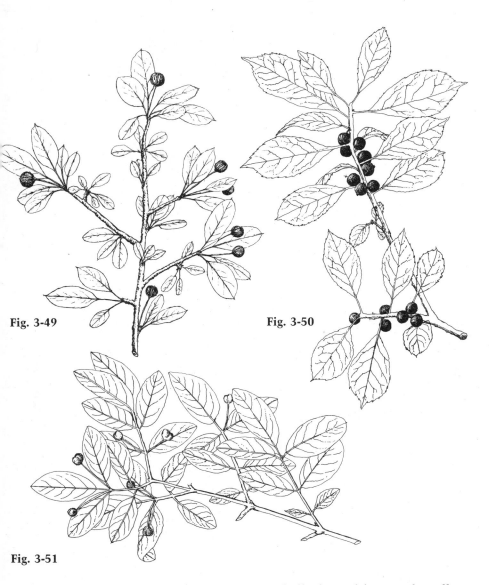

Fig. 3-49. *Nemopanthus mucronatus*, mountain holly, has red fruits and small, more-or-less entire leaves ending in a minute bristle. It belongs to the highbush zone peripheral to bog mats. **Fig. 3-50.** *Ilex verticillata*, Michigan holly, also called winterberry, has bright red berries clustered along its branches. The berries make a glorious display after leaf fall and persist well into the winter. The plants grow in shrubby swamps and laggs. **Fig. 3-51.** *Lonicera oblongifolia*, fly honeysuckle, is a small shrub with opposite leaves and long-stalked, red berries joined in pairs. It grows in rich fens.

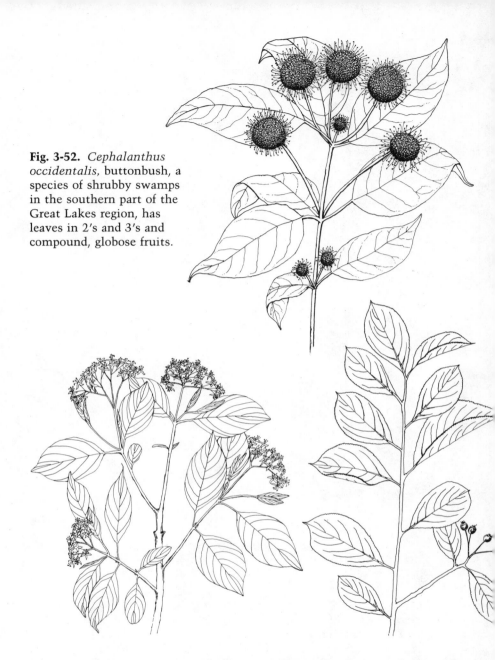

Fig. 3-52. *Cephalanthus occidentalis*, buttonbush, a species of shrubby swamps in the southern part of the Great Lakes region, has leaves in 2's and 3's and compound, globose fruits.

Fig. 3-53. *Cornus stolonifera*, red osier, has red stems, opposite leaves with lateral veins curved toward the tip, and white fruits. It grows in shrubby swamps and southward, at least, in shrubby zones marginal to bog mats.

Fig. 3-54. *Aronia prunifolia*, chokeberry, is a member of high-bush associations. It has glandular-serrate leaf margins and also a scattering of glands on the base of the midrib, on the upper surface. The fruits are dark purple to blackish.

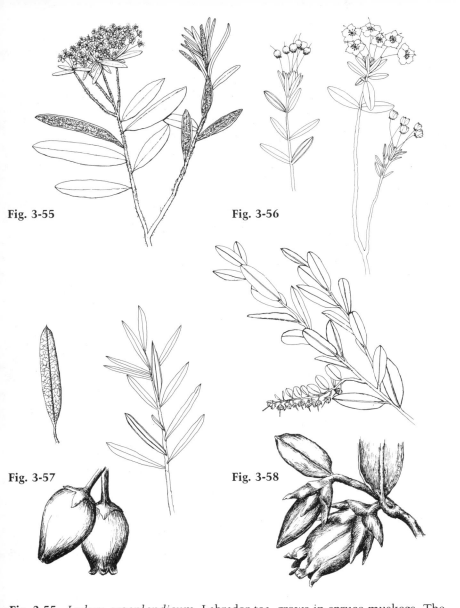

Fig. 3-55. *Ledum groenlandicum*, Labrador tea, grows in spruce muskegs. The leaves are brown-woolly underneath and the flowers white. **Fig. 3-56.** *Kalmia polifolia*, bog laurel, has shiny-green opposite leaves and pink flowers. It grows in relatively dry niches, most commonly in the hummock-hollow, low-shrub zone of bogs. **Fig. 3-57.** *Andromeda glaucophylla*, bog rosemary, has blue-gray leaves, alternate in arrangement, with white undersurfaces and revolute margins. The flowers are white and bell-shaped. *Andromeda* grows in wetter portions of floating mats, especially in poor fens. **Fig. 3-58.** *Chamaedaphne calyculata*, leatherleaf, has leaves with scurfy brown undersurfaces. The white bell-shaped flowers are borne on terminal branches with small leaves that drop off during the first year. The other leaves fall gradually over a number of seasons. The flowering branches assume a lateral position before they too drop off. *Chamaedaphne* dominates the low-shrub zone of bogs but also grows in poor fens and intermediate fens.

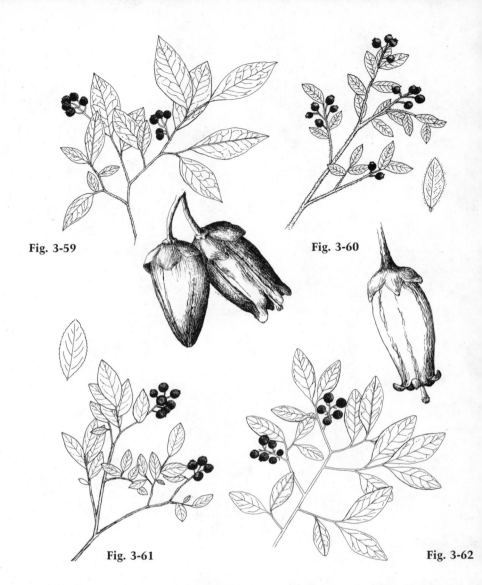

Fig. 3-59. *Vaccinium corymbosum*, high-bush blueberry, is well represented at the periphery of peatland mats in the southern part of the Great Lakes region. **Fig. 3-60.** *Vaccinium myrtilloides*, blueberry, grows in open bog mats, often successional to fire, and has pubescent leaves and branches. **Fig. 3-61.** *Vaccinium angustifolium*, blueberry, is a low shrub of open bogs, often in places burned over. **Fig. 3-62.** *Gaylussacia baccata*, huckleberry, grows in relatively dry peatland sites, in spruce muskeg, for example. The leaves are dotted on both surfaces by resin glands yellow-staining a paper if the leaves are rubbed against it. Minute bracts produced among the flowers and fruits, though deciduous, help to distinguish huckleberries from blueberries, but the basic difference is that huckleberries have ovaries with 10 cells and hard seeds, whereas blueberries have 4–5-celled ovaries and soft seeds.

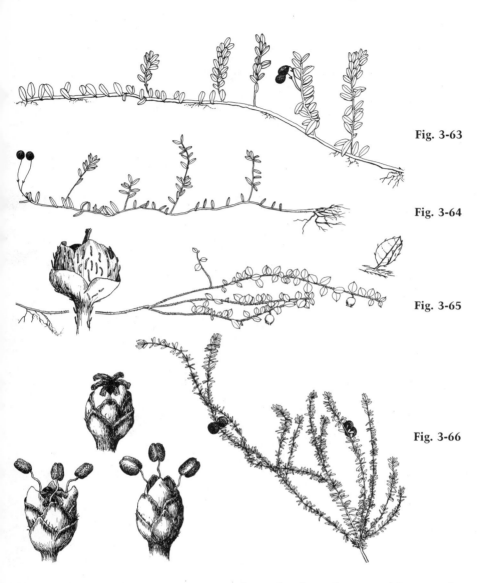

Fig. 3-63. *Vaccinium macrocarpon*, cranberry, has larger leaves and fruits, and its pedicels have bracts above the middle. **Fig. 3-64.** *Vaccinium oxycoccos*, small cranberry, has bracts at or below the middle of its pedicels. It is characteristic of the poor fen (*Sphagnum* lawn) community, whereas *V. macrocarpon* is more common in somewhat less acid habitats in the sedge fen that precedes the development of poor fen. **Fig. 3-65.** *Gaultheria hispidula*, creeping snowberry, has stems and undersurfaces of leaves with woody-seeming hairs and white fruits. **Fig. 3-66.** *Empetrum nigrum*, black crowberry, is a small, woody plant of acid heaths at northern latitudes but occurs, rarely, southward in rich fens. Its black berries contrast with red ones of cranberries. Its flowers can be male, female, or mixed on the same plants.

Fig. 3-67. *Carex lacustris* **Fig. 3-68.** *C. aquatilis.* **Fig. 3-69.** *C. comosa*

Fig. 3-70. *Carex pauciflora.* **Fig. 3-71.** *C. oligosperma.* **Fig. 3-72.** *C. stricta* is a species of marshes and meadows that, in the past, was harvested as "marsh hay." In addition to feeding and bedding livestock, marsh hay was also used for insulation, to keep ice through the summer and keep Milwaukee beer cold until it reached Chicago. As "marsh grass," the hay is now sold as a garden mulch for keeping down weeds. **Fig. 3-73.** *C. rostrata.*

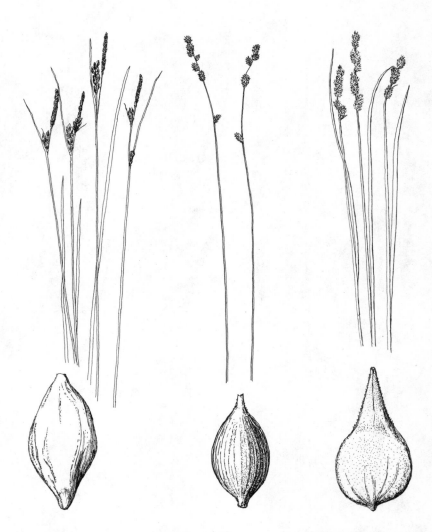

Fig. 3-74. *Carex livida*

Fig. 3-75. *C. canescens* has gray-green or glaucous foliage, and its lower 1 or 2 female spikes are well spaced.

Fig. 3-76. *C. diandra* has pale whitish leaf sheaths dotted with purple and crowded spikes.

Fig. 3-77. *Carex trisperma* has few perigynia per spike and a long, slender bract subtending the lowest spike.

Fig. 3-78. *C. disperma* lacks a subtending bract but otherwise resembles *C. trisperma*.

Fig. 3-79. *C. lasiocarpa* has fuzzy perigynia. It is the main pioneer mat former at the margin of alkaline lakes.

Fig. 3-80. *C. paupercula* grows in tufts (not from sprouted culms), and its leaves are green. The bracts of the female spikes are broader and the male spikes are shorter than in *C. limosa*.

Fig. 3-81. The female inflorescence of *C. limosa* is bicolored because of bronze bracts that allow the green perigynia to show. The culms elongate and fall over after flowering, and young plants sprout at their nodes. The leaves are a pale bluish green.

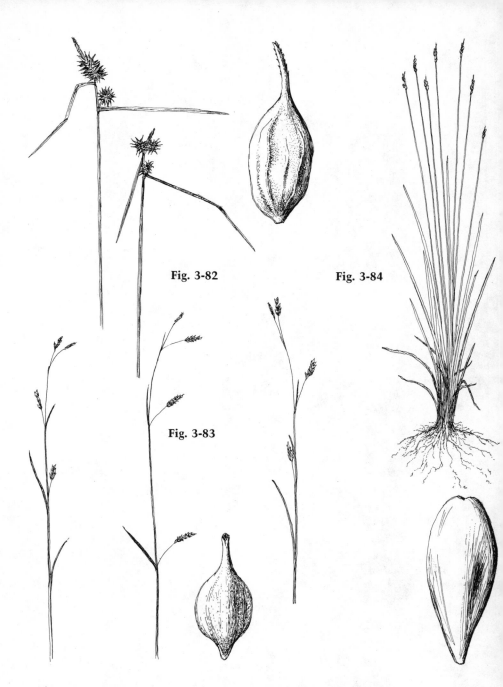

Fig. 3-82. *Carex flava*, of mineral-rich habitats, has wide-spreading bracts below yellow inflorescences. **Fig. 3-83.** *C. capillaris* has lowermost spikes that droop on slender pedicels. **Fig. 3-84.** *C. leptalea* has small spikes and broad, blunt perigynia.

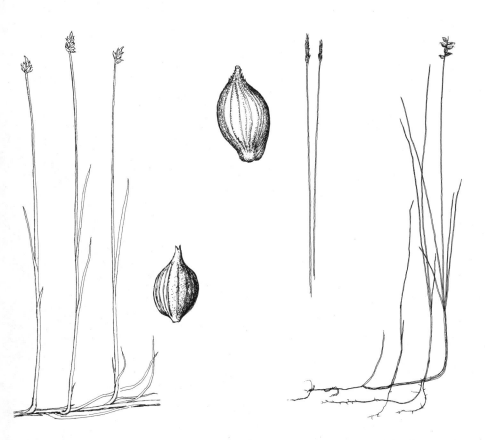

Fig. 3-85. *Carex chordorrhiza*, like *C. gynocrates* and *C. limosa*, sprouts new plants from the nodes of fallen culms. The spikes, staminate at their tips, are crowded into what appears to be a single inflorescence.

Fig. 3-86. *C. gynocrates* is almost always dioecious, but rarely (in the Great Lakes region) some plants have staminate inflorescences subtended by a few perigynia. (In some parts of the broad species range, the female portion of the spike may be better developed.)

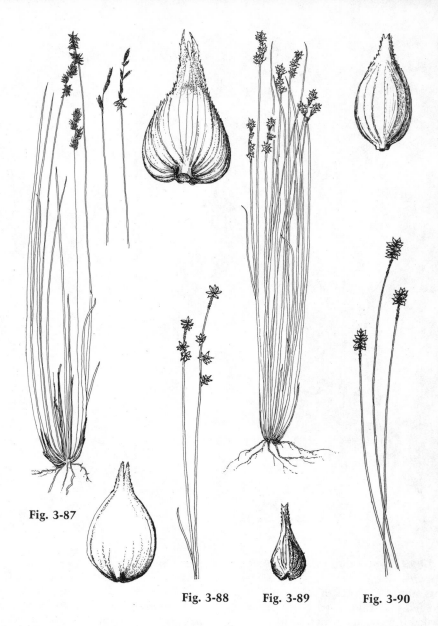

Fig. 3-87. *Carex sterilis*, restricted to wet, calcareous habitats (rich fens), may be dioecious or monoecious, in which latter case the male spikes are subtended by a group of perigynia. **Fig. 3-88.** *C. interior* has spikes that are female except at the base. This condition is quite conspicuous in the topmost spike where the slender male portion is quite elongate. **Fig. 3-89.** *C. cephalantha*, a plant of acid habitats, has terminal spikes somewhat club-shaped because the male inflorescence subtends a broader group of perigynia. **Fig. 3-90.** *C. exilis* has solitary, terminal spikes that are female above, male below. (The perigynia are shown out of proportion on this plate of illustrations.)

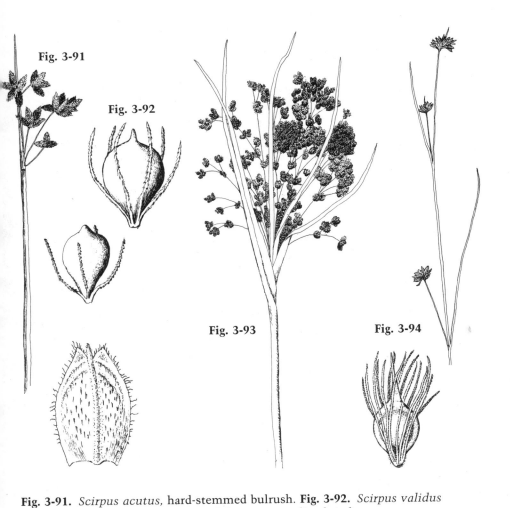

Fig. 3-91. *Scirpus acutus,* hard-stemmed bulrush. **Fig. 3-92.** *Scirpus validus* differs in details of the achenes and their surrounding bristles.
Fig. 3-93. *Scirpus cyperinus,* wool grass, is common in open, wet laggs.
Fig. 3-94. *Rhynchospora alba,* white beak rush, is especially characteristic of poor fens.

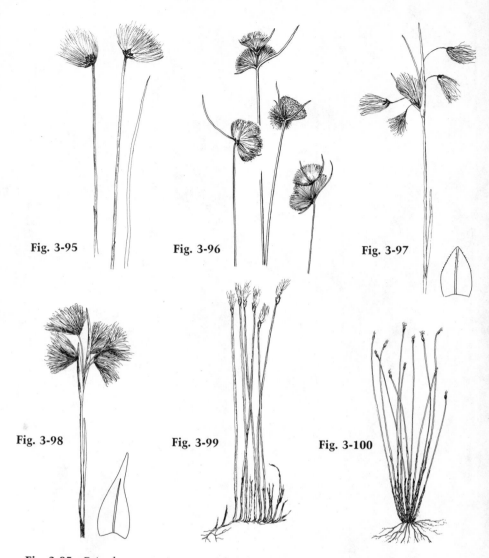

Fig. 3-95. *Eriophorum spissum*, with solitary white, cottony spikelets.
Fig. 3-96. *Eriophorum virginicum*, tawny cotton grass, has a few spikelets on short pedicels crowded together in an erect group conspicuously subtended by 2 or sometimes 3 relatively long bracts (involucral leaves). **Fig. 3-97.** *Eriophorum viridi-carinatum* has groups of white spikelets on long pedicels. The spikelets are surrounded by small scales midveined to their tips. **Fig. 3-98.** *Eriophorum angustifolium* has loose groups of white spikelets on long pedicels. The small scales surrounding each spikelet have midveins ending short of the apex. **Fig. 3-99.** *Scirpus hudsonianus*, alpine cotton grass, a rush of rich fens, has white perianth bristles that give some resemblance to the spikelets of *Eriophorum*. **Fig. 3-100.** *Scirpus cespitosus* grows in rich fens in the upper Great Lakes region (but elsewhere sometimes in quite acid habitats). The plants form large, dense tufts, often windblown in appearance. The inflorescences are not in the least cottony.

Fig. 3-101. Fig. 3-102. Fig. 3-103.

Fig. 3-101. *Xyris montana*, yellow star grass, is a small plant of poor fens in the upper Great Lakes region. Farther south the similar *X. diffusa* is common.
Fig. 3-102. *Cladium mariscoides* is a tall sedge that forms mats in particularly wet and mineral-rich sites, sometimes in association with *Scirpus acutus*.
Fig. 3-103. *Dulichium arundinaceum*, the 3-way sedge, is common in wet, pioneering sedge communities. The leaf blades are conspicuously spreading in 3 rows.

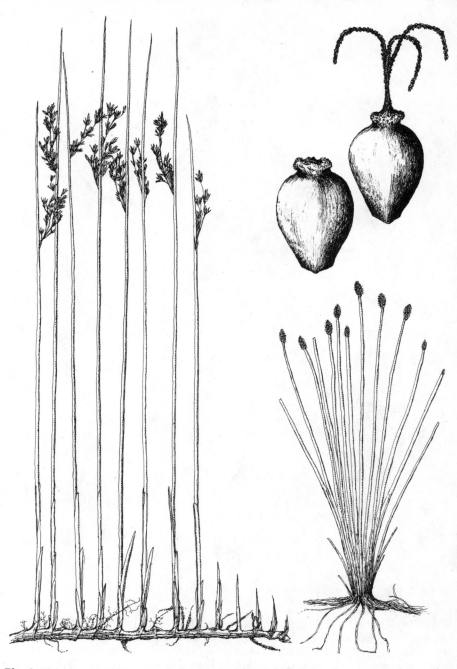

Fig. 3-104. *Juncus balticus.* This rush grows evenly spaced in straight rows on marly wet soil of rich fens (and also on lakeside sand).

Fig. 3-105. *Eleocharis ellipticus* grows in scattered clumps on exposed lake mud which is later invaded by *Carex lasiocarpa*. The plant consists of a clump of culms, most of them ending in a small, ellipsoidal spikelet. There are no leaves. (The various species differ in size but look pretty much alike.)

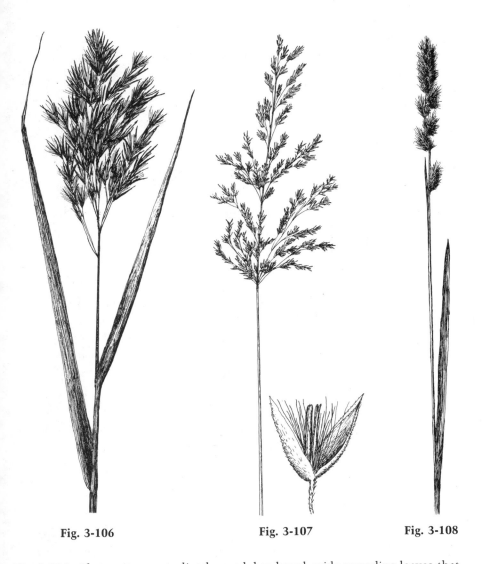

Fig. 3-106.   Fig. 3-107.   Fig. 3-108.

Fig. 3-106. *Phragmites australis*, the reed, has broad, wide-spreading leaves that resemble pennants. (The leaves have been traditionally used in England for thatching roofs.) The tall plants grow in rich fens, especially in transition to *Thuja* swamps. **Fig. 3-107.** *Calamagrostis canadensis*, blue joint, is found in rich and intermediate fens, in sedge mats and at wet margins of open laggs. **Fig. 3-108.** *Muhlenbergia glomerata*, marsh wild timothy, grows in wet, mineral-rich places such as intermediate fens.

Fig. 3-109. *Glyceria canadensis*, rattlesnake grass, is found in wet, mineral-rich sites, in laggs, for example.

Fig. 3-110. *G. striata*, fowl manna grass, grows in laggs and rich fens transitional to *Thuja* swamps.

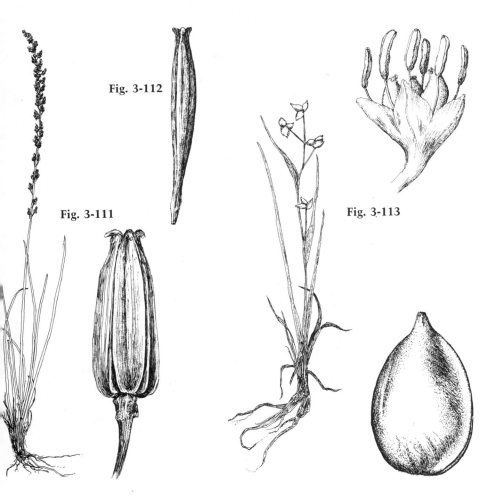

Fig. 3-111. *Triglochin maritimum*, arrow grass, is common in wet, calcareous habitats, in rich fens and at margins of beach pools in the vicinity of the Great Lakes. **Fig. 3-112.** *Triglochin palustre* has much narrower fruits with 3 carpels rather than 6. **Fig. 3-113.** *Scheuchzeria palustris* grows in very wet sites in *Sphagnum* lawns. The leaf tips have a conspicuous white pore.

Fig. 3-114. *Tofieldia glutinosa*, false asphodel, a modest plant of marly fens, has sticky-glandular flower stalks. **Fig. 3-115.** *Parnassia glauca*, grass of Parnassus, also of rich fens, has solitary, rather obviously veined, white flowers. **Fig. 3-116.** *Pyrola asarifolia*, shin leaf, has pale pink flowers that mature from the base of the inflorescence upward.

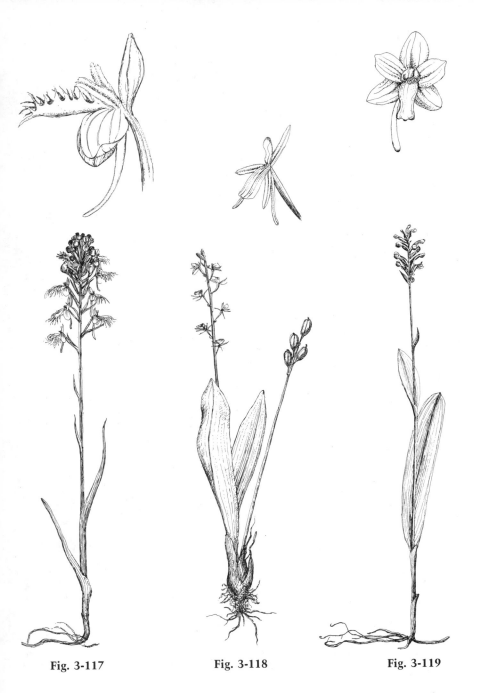

Fig. 3-117.   Fig. 3-118.   Fig. 3-119

Fig. 3-117. *Habenaria blephariglottis*, white fringed orchid, grows in poor fens. It is somewhat rare in northern Michigan but common in some other parts of the upper Great Lakes region. **Fig. 3-118.** *Liparis loeselii*, the green twayblade or fen orchid, is hidden away among sedges in intermediate fens. The flowers are small and greenish- or yellowish-white in loose racemes.
Fig. 3-119. *Habenaria clavellata*, club-spur orchid, has small, greenish or yellowish flowers. It can be expected in poor fens.

Fig. 3-120.   Fig. 3-121.   Fig. 3-122.

Fig. 3-120. *Arethusa bulbosa*, dragon's mouth, has bluish pink flowers with the lower lip streaked with purple and yellow. It can be expected in poor fens. **Fig. 3-121.** *Pogonia ophioglossoides*, rose pogonia, has pale, rose-pink flowers. The lower lip is fringed at the margin, veined with red, and bearded with yellow hairs. It is especially common in poor fens. **Fig. 3-122.** *Calopogon tuberosus*, the grass pink, is common in intermediate and poor fens. Its flowers are deep pink and have the lip uppermost, in contrast to the position in *Arethusa* and *Pogonia*.

Fig. 3-123. *Cypripedium acaule*, the pink (or stemless) lady slipper, grows in a variety of habitats, among them spruce muskegs. It has two basal leaves. The sepals are relatively short and yellow-green to brownish, as are also the lateral petals. The moccasin-like lip is pink.

Fig. 3-124. *C. calceolus*, the yellow lady slipper, is a denizen of sedgy transitions from rich fen to *Thuja* swamp. It has a leafy stem. The moccasin is yellow, and the sepals are purplish brown, relatively long, and twisted.

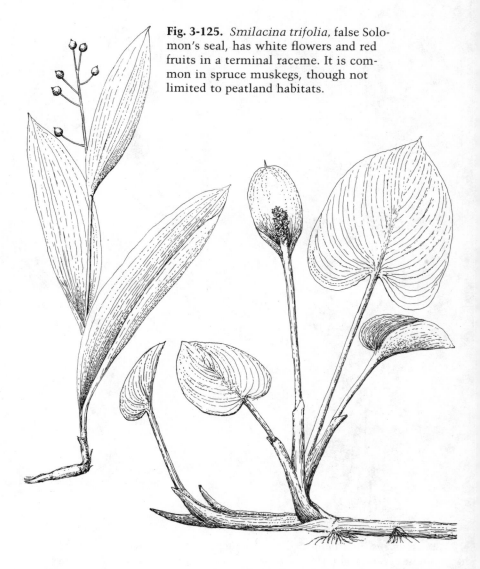

Fig. 3-125. *Smilacina trifolia*, false Solomon's seal, has white flowers and red fruits in a terminal raceme. It is common in spruce muskegs, though not limited to peatland habitats.

Fig. 3-126. *Calla palustris*, the wild calla, is common in alder swamps occupying wet laggs. The white, expanded spathe subtends an elongated spadix covered with small flowers and eventually bright red berries.

Fig. 3-127. *Menyanthes trifoliata*, the buckbean, grows in wet, mineral-rich habitats, especially in the sedge mat bordering alkaline lakes.

Fig. 3-128. *Iris versicolor*, the wild blue flag, grows among sedges in wet, mineral-rich places, often in laggs and pioneering sedge mats.

Chapter 4

Nutrients in Scant Supply

> Never can the priest so describe hell, because it is no worse. Never have poets been able to picture Styx so foul, since that is no fouler.
> —Linnaeus

The quiet beauty of a bog conceals a contentious environment, cold, wet, deficient in amenities, but not so hellish as Linnaeus described the mires of Lapland to be. A bog is a stressed ecosystem where few plants survive. The strategies by which they eke out an existence include, above all, a frugal use of nutrients. Death and decay release nutrients to be used again, to be cycled and recycled. Bacteria, fungi, and invertebrates are involved in the release, and a complete breakdown requires the action of many decomposers. But decomposers have only limited effectiveness in cold, wet, sour peat, and slow decay compounds the deficiencies of a starved ecosystem.

Aeration favors decomposition and nutrient release in swamps, marshes, meadows, and fens. But in wetlands of reduced aeration, in bogs, microbial action is diminished and especially at levels of peat where the products of anaerobic decay, such as alcohol and a variety of organic acids, act as inhibitors.

> Nature ... is a friend you will never lose until death—and even when you die, you disappear into nature.
> —Tolstoy

Fens are rich in nutrients, or eutrophic, whereas bogs are nutrient poor, or oligotrophic. The terms, originally applied in that sense to peatlands, later came to refer to lake-water concentrations in phosphorus, nitrogen, and calcium and to productivity. The more eutrophic the lake the greater the algal productivity and the more turbid the water. Still later, the terms came to express oxygen content. Deep lakes with dark, cold underlevels deficient in oxygen and unproductive in terms of algae are oligotrophic, but shallow lakes relatively productive because of high oxygen content are eutrophic. In any sense, whether applied to peatlands or lake waters, eutrophy and oligotrophy bear connotations of productivity and species diversity.

Bogs depend on the atmosphere for nutrients. Near the ocean precipitation is enriched by salt spray. Volcanic and industrial pollutants, agricultural dust, and gases resulting from biological conversions also contribute nutrition. Nitrogen is brought in by rain and snow and airborne organic detritus. The Minnesota peatlands get a generous supply of calcium-rich dust from the Dakota wheat fields. In the broad expanse of Minnesota peatlands, black spruce collects dust, but bog shrubs of uniformly low stature trap dust less effectively. Small bogs, in depressions, are sheltered from dust-laden winds, but their spruce trees do intercept and concentrate rain-carried minerals. The amount of calcium used in Finnish bogs is just about equal to the amount brought in from the atmosphere, but for potassium, phosphorus, and nitrogen the amounts used exceed the amounts deposited from the atmosphere (180). It thus appears that those anion nutrients need to be recycled. The element most actively recycled and taken up by bog plants is potassium, followed in order by phosphorus, nitrogen, magnesium, and calcium.

Snowmelt and rainfall wash nutrients into laggs. A redistribution into less nourished zones is facilitated by water track drainage and criss-crossing deer trails (281) (fig. 4-1). Nitrogen fixed in the root nodules of alder and sweet gale—*Alnus* and *Myrica*—can be traced along such drains from eutrophic laggs and sedge mats into oligotrophic bogs, and nutritive increases can be detected by vegetational responses along the drains. In large peatlands in Minnesota, a slight doming of the bog surface allows for water movement revealed by radiating stands of spruce. Aeration associated with moving water increases decay and nutrient release in the same patterns of radiation. The leaves of alders and tamaracks growing in base-rich zones are high in nutrients. Their annual fall and slow decay renew the ecosystem, as does the leaf litter that laggs receive from adjacent woodlands.

At least sixteen elements are needed for plant growth. The *macronutrients*, those needed in quantity, are carbon, hydrogen, oxygen, nitrogen, phosphorus, potassium, sulfur, calcium, and magnesium. Equally important but needed in trace amounts are the *micronutrients*, iron, manganese, boron, zinc, copper, molybdenum, and chlorine. Carbon, hydrogen, oxygen, and nitrogen are derived, ultimately at least, from air and water. The usable forms of these essential elements are free oxygen and oxides of carbon and hydrogen. The other nutrients come from weathered rock and soil or decayed organic matter.

Nutrient Availability

The mineral constituents of water-soaked peat are locked away by conditions unfavorable for decay. To be available nutrients need to be combined in soluble forms. The suitable combinations of ions vary with pH. A considerable range of pH, or hydrogen-ion concentration, can be demonstrated in peatlands. The degrees of concentration are expressed on a logarithmic scale of 1 to 14, with acid values below and alkaline values above 7. A value of 3 is 10 times as acid as 4, 100 times as acid as 5, and so on. The values found in peatlands range from 3 in bogs to 8 in fens. This means, in a sense, that bogs may be as much as 100,000 times as acid as fens!

Seasonal variations in pH (related to a greater ion exchange during the growth period) and dilution brought on by rain add complexity to the bog habitat. The acidity of hummocks cannot be tested by a standing water method. Probing with a pH meter into various levels of a hummock or testing squeezings is reasonably reliable if some attempt is made to equalize the temperature of the buffer solution with that of the water to be tested. This can be done by making squeezings in advance of testing and allowing buffer and sample to become temperature adjusted or by cooling the buffer in the hummock before testing.

The ability of roots to take up nutrients is impaired at pH levels less than 5, and temperatures too low for metabolic processes, as well as the lack of oxygen, further reduce uptake. Macronutrients are generally available for uptake at pH 4–9, but some of the micronutrients show much narrower ranges of availability. The total content of most minerals decreases along a fen-to-bog gradient. Most of them have their greatest availability below pH 7.5, but restricted availability at 4.5–5. A minimum of 3 is required. Some trace elements, like molybdenum, are poorly available in acid waters. Nitrogen and sulfur are readily available throughout the fen range, actually from pH 5 to 9. Manganese and phosphorus, on the other hand, may be more available in the acidity of bogs owing to an increased solubility of manganese and a reduced content of metallic ions with which phosphorus can combine in insoluble forms.

Generally speaking, the best pH for nutrient availability in fen peats is around 5.5–5.8 and in *Sphagnum* peats about 5 (204). This means that a poor fen, or wet *Sphagnum* lawn, makes good use of nutrients in support of the productivity needed for a buildup in peat sufficient for a fen-to-bog transformation. A greater availability shows up in the mineral content of fen plants, about twice that of bog plants.

Nutrient Cycling

> Life—human life included—is the outcome of an elaborate organization based on trivial ingredients and ordinary forces.
>
> —Palade

Nutrient cycling depends on decay. Complete decay represents the combined activities of many kinds of bacteria and fungi. Some decomposers are aerobic, and some are not. Anaerobic decomposition is limited and incomplete, and bacteria involved in nitrification and nitrogen fixation are, for all practical purposes, absent from waterlogged peat.

Many nutrients occur in plant cells as inorganic salts or loosely held components of organic molecules. Organic potassium and phosphorus can be readily freed by decay, but nitrogen and sulfur may not be. Peat is especially poor in potassium, phosphorus, and usable nitrogen. Leaf assays of bog ericads show those nutrients at levels inadequate for plants of other habitats, but the amounts of iron and manganese are much more than adequate. A high content in manganese in bog plant tissues suggests that tolerance to that mineral, generally toxic in all but slight amounts, bestows a competitive advantage.

Bog plants manufacture carbohydrates with efficiency in spite of nutrient shortages. The evergreen leaves of some of the ericad shrubs conserve the energy that other plants use in the annual renewal of leaves. The persistence of leaves for as much as four years provides longer seasons for photosynthesis. The leaves of *Chamaedaphne* remain on the plant less than two full years, but their gradual fall allows some continuity of photosynthesis and nutrient release by decay.

The accumulation of peat seriously limits nutrient availability. It is said that a large share of the nutritional requirements of Canadian spruce forests, including muskegs, could be met by the decomposition of mosses that make up the ground cover and the peaty substrate. The rate of decomposition varies with the kind of organic matter. Sugars, starches, and proteins go first, followed in order by hemicellulose, cellulose, lignin, and waxes. Hemicelluloses are polysaccharides of some 50–100 sugar units, and they are fairly susceptible to decay. Cellulose consists of about 1,000–10,000 sugar units and is significantly more decay resistant. Lignin, consisting of a number of aromatic units linked by aliphatic side chains, is extremely resistant to decay, but it is eventually transformed into humic substances. Lignin is an ill-defined class of compounds associated with the secondary walls of vascular and support tissues of higher plants. Sphagnol is a decay-resistant, ligninlike substance held in the walls of *Sphagnum.*

Nitrogen

Nitrogen is abundant in the atmosphere in its inert molecular form. For plant use it needs to be in nitrate or ammonium forms, both readily soluble and easily lost by leaching. Peat contains rather substantial amounts of nitrogen (about 0.5 percent in *Sphagnum* peat and 3 percent in sedge peat on an oven-dry basis), but seldom is more than 5 percent of it available for uptake. About 40 percent of the total nitrogen is held in organic compounds that require chemical action for conversion to soluble forms. Such conversions depend on decay organisms that are, however, not very active under conditions of acidity. In drains and soaks where surface waters move through bogs, there are more oxygen, more minerals, and less acidity. Increased microbial activity in such areas increases nutrient cycling and the availability of nutrients, including nitrogen. The bacterial activity reduces the amount of peat accumulated in water tracks and thus perpetuates them as a means of nutrient distribution. Both total and available nitrogen decrease in amount from fen to bog conditions. If the pH is less than 5, as in ombrotrophic bogs, nitrate release by protein decay is much decreased. In anaerobic peat where decomposition is virtually lacking, organic nitrogen accumulates, insoluble, unavailable. So it is that the total nitrogen in bog peat is high, but availability is low.

Decomposers can use organic nitrogen but prefer ammonium and nitrate ions in the soil solution. They are thus in competition with green plants that can take up and use only these soluble, inorganic forms. The microorganisms are, however, short-lived, and their nitrogen is quickly cycled.

Decay organisms are adversely affected by water-soaking. Because of its wetness, peat is devoid of oxygen about 8 inches below the bog surface. A clean copper wire forced into the peat remains bright above but blackens by sulfide corrosion in the anaerobic reduction zone (which can also be detected by the rotten-egg smell of hydrogen sulfide). Acidity, as well as mineral content, aeration, and temperature, determines differences in microbial populations and activities. Bacteria predominate in acid, anaerobic peat and are also active as aerobes at circumneutral pH values. However, they are much fewer in number and kind in the acidity of low-shrub mats and spruce-tamarack stands than in cedar swamps. Fungi and actinomycetes are active in acid, aerobic environments.

In the presence of oxygen, the chief products of protein decay are nitrate, ammonium, and sulfate ions, water, and carbon dioxide. But in the absence of oxygen, proteins are less quickly and less completely degraded to ammonia, hydrogen sulfide, and a variety of strong-smelling organic compounds in addition to amino acids and carbon dioxide. Anaerobic bacteria break down cellulose by fermentation, yielding ethyl alcohol plus lactic, acetic, succinic, and formic acids. A secondary group of bacteria use these acids to generate methane (CH_4), or marsh gas. In the case of acetic acid, methane can be produced in two tandem reactions:

$$CH_3COOH \rightarrow CH_4 + CO_2$$
$$CO_2 + 8H \rightarrow CH_4 + 2H_2O$$

Methane is highly reduced and will readily ignite on contact with air. It is likely that the will-o'-the wisp or jack-o'-lantern, as the Irish call ghostly flashes of light passing over peatlands, is a result of the oxidation of methane and also of the highly reduced gas phosphine (PH_3), which is spontaneously flammable owing to the presence of the hydride form (P_2H_4). The phenomenon apparently can occur, but it no doubt belongs largely in the realm of fantasy and superstition.

Numerous bacteria take part in nitrogenous decay, but the process begins with those capable of external digestion. Protein molecules are too large to pass through cell membranes, but products of protein digestion, such as amino acids, can be absorbed and further altered by other agents of decay.

Green plants favor nitrate nitrogen under acid and ammonium nitrogen under less acid conditions. The poor use of ammonium nitrogen in acid environments is linked to a deficiency of salt-forming anions and hence an escape from solution as gaseous ammonia. Nitrites can be taken up by roots, but they rarely accumulate in the soil because of solubility and microbial action. In any case, under conditions of acidity, nitrites are relatively toxic. Most amino acids are toxic.

Nitrogen release is related to the carbon:nitrogen ratio. Carbon is plentiful in plant remains, nitrogen much less so. Microbial activity is inhibited by a high C:N ratio. The reason is obvious. The microbes need nitrogen in order to live, multiply, and decompose carbohydrates effectively. Litter with a good supply of nitrogen decomposes quickly, but *Sphagnum* peat is very low in nitrogen and very slow to decay. The C:N ratio in sedge peats is less than 20:1, while that in *Sphagnum* peat is sometimes as much as 60:1. The lower ratios in sedge peats have to do with aeration and its effect on decomposition. Oxygen deficiency and acidity inhibit decay in *Sphagnum* peat, and low concentrations of nitrogen held by *Sphagnum* plants, owing to a large component of empty nonliving cells, contribute to high C:N ratios in bog peat. Under conditions of oxygen availability, fungi are adept at breaking proteins down to amino acids, after which other aerobes join in the decay process. Farther down, decomposition depends on anaerobic bacteria, and the rate of protein decay is greatly reduced. The result is that the anaerobic decomposers of cellulose and lignin are deprived of the nitrogen they need. Nitrogen-freeing microbes, on the other hand, depend on simple sugars in the soil for their energy source, and they have to compete for a small and transient supply. The result is a deadlock. Both sugar and nitrogen remain essentially unavailable, and little decomposition of complex carbohydrates or proteins takes place.

Many soil bacteria, aerobic and anaerobic, are capable of *ammonification*, that is, the breakdown of amino acids and release of ammonia:

$$\text{"CHONS"} + O_2 \rightarrow CO_2 + NH_4OH + SO_4^{--} + \text{energy}$$
$$NH_4OH \rightarrow NH_3 + H_2O$$

(Of course, not all amino acids contain sulfur.)

The ammonia then undergoes *nitrification,* or oxidation, in a two-step process dependent on two kinds of aerobic bacteria:

$$2NH_3 + 3O_2 \rightarrow 2NO_2^- + 2H^+ + 2H_2O + \text{energy}$$
$$2NO_2^- + O_2 \rightarrow 2NO_3^- + \text{energy}$$

The bacteria involved in nitrification use the energy of oxidation in the synthesis of carbohydrates just as green plants and some bacteria use light energy. They are chemosynthetic rather than photosynthetic.

If the pH falls below 5, there is a marked decrease in nitrification. In oxygen-poor soils some bacteria reduce nitrate to nitrite and free nitrogen as an oxygen source for the combustion of carbohydrates. Considerable *denitrification* takes place in waterlogged peat. Ammonia produced by decay or brought in with the rain is nitrified in the upper few centimeters of the bog surface, but the nitrate so formed is denitrified as it soaks down into anaerobic peat. Most peatlands are subject to alternating periods of wetting and drying. Nitrate formed during the dry, aerobic periods is quickly denitrified on wetting.

Nitrogen can be returned to use by *nitrogen fixation.* Some atmospheric nitrogen is oxidized by electrical discharge, or lightning, and industrial pollutants also contribute oxides of nitrogen to the atmosphere. Ammonia is released to the air by the combustion of coal and oil, and some escapes from the soil as a result of protein decay. Rain and snow bring these nitrogen compounds back to earth. But much more important as a source of usable nitrogen is biotic fixation, by free-living and symbiotic bacteria (including actinomycetes) and blue-green algae.

Nitrogen fixers reduce nitrogen to ammonia, which may be excreted or used to synthesize organic nitrogen compounds. Both free-living and symbiotic nitrogen fixers leak ammonia into the soil, but most of it goes into organic nitrogen that becomes available to other organisms only after death and decay take place. The nitrogen-fixing reaction— $2N_2 + 6H_2O \rightarrow 4NH_3 + 3O_2$ —requires the enzymatic action of nitrogenase, which has in its molecular structure iron and molybdenum and requires copper for activation. These minerals are present only in low concentrations in acid environments. Nitrogenase is limited in its action to low-oxygen situations.

Twenty-eight genera and some 60 species of blue-green algae fix nitrogen. (Some of them are symbiotic with lichen-forming fungi.) They achieve low oxygen tensions by having heavy gelatinous sheaths, growing in clumps, or localizing anaerobic metabolism in specialized cells, or heterocysts. Low oxygen concentrations are maintained in heterocysts by the virtual absence of oxygen-releasing photosynthesis and a high rate of oxygen-using respiration. (The heterocysts use photosystem I to supply energy for nitrogen fixation but lack photosystem II—the pathway that involves oxygen.) Blue-greens characterize nutrient-rich habitats, yet many of them occur in bogs, in association with *Sphagnum,* among close-packed leaves and inside water-holding cells. Blue-greens do best at pH values above 5, but cation exchange at the cell surfaces of *Sphagnum* may create microenvironments buffered against acid extremes. Algal associates of *Sphagnum* include euglenoids, desmids, diatoms, and heterocystous blue-greens, some of the latter "proved" by acetylene reduction to fix nitrogen. Because of low phosphate concentrations, however, blue-green algae are relatively unimportant as nitrogen fixers in bogs.

In mineral-rich peatlands, *Alnus* and *Myrica* form root nodules that provide anaerobic sites as well as carbohydrates for the use of actinomycete nitrogen fixers. Fixed nitrogen can be traced into bogs from fen sites occupied by *Alnus* and *Myrica,* and nodules on *Shepherdia, Comptonia,* and *Ceanothus* roots may also contribute nitrogen to laggs through seepage from upland slopes. Most legumes have root nodules occupied by nitrogen-fixing bacteria. Legumes are common on mineral soils of nitrogen deficiency but in North America, at least, do not occur on peat. *Myrica* fixes nitrogen in a quantity comparable to the legumes and can do so at greater acidity (152). Because nitrification and nitrogen fixation are essentially lacking at bog pH levels, the nitrogen available to bog plants is provided, mainly, though scantily, by precipitation and seasonal influx through soaks and animal trails.

Nitrogen fixation is most active in poor, upland soils, yet nitrogenase activity has been detected in a whole range of fens (including poor fens) from the High Arctic south to 47° N in Germany, but none was found in bogs. Some of this activity may be attributed to the bacterium *Azotobacter,* which has frequently been found in peatlands, and some no doubt to blue-green algae. In central Europe, vascular plants with nitrogen-fixing root nodules, members of the Leguminosae and the genus *Alnus*, are restricted to fens. *Myrica gale* grows in fens in western continental Europe, but in the west of Scotland and Ireland it is abundant in ombrotrophic peatlands (152).

Phosphorus

Phosphorus does not occur free in nature. Unstable in the presence of oxygen, it occurs in oxidized forms, generally phosphates dissolved out of rock or soil by water, probably aided by carbonic acid. Most phosphorus is lost by drainage, and so continued weathering is essential. Phosphate ions are only scantily represented in rainwater or dry fallout. Phosphorus occurs in living cells as organic phosphates.

Phosphorus limits the growth of bog plants because so little is available. The ratio of phosphorus to other elements held in organic combinations tends to be considerably higher than that in the inorganic environment. In mineral soils, bacteria rapidly convert organic to inorganic phosphates that are readily absorbed and concentrated by green plants. But in bogs only a meager supply comes from organic phosphates because of slow decay. Phosphorus also occurs in peat in mineral compounds, mainly as phosphates of aluminum, iron, and calcium. In aerated fen peat these compounds are mostly insoluble, but in waterlogged, acid peat more soluble compounds are formed. Thus, total phosphorus decreases in amount from fen to bog conditions, but the amount available increases.

The Ericaceae, in or out of the bog habitat, cope well with phosphorus-poor soils. In many of them, the long retention of leaves and seasonal transfer of nutrients, as well as mycorrhizal relationships, may contribute to their success in marginal habitats. Even though phosphorus is scarce in bog soils, plants may concentrate inside their cells adequate amounts as a result of active uptake from dilute solutions (that is, energy-using movement against a concentration gradient).

Potassium

Rain and snow are low in potassium content, and hence bogs are deficient in that element. Weathered from soil and rock, potassium easily combines with other products of weathering and binds with clay particles, by adsorption and chelation (that is, by chemical binding between atoms of "clawlike" chelator molecules, frequently between atoms of nitrogen, oxygen, or sulfur). Potassium is rapidly released by decay and rapidly lost by leaching. Organic soils are low in potassium because they have little mineral matter capable of holding potassium and releasing it gradually, and in peat roots do not penetrate to the water-soaked level where complexing by humic colloids is possible. Also, being monovalent, potassium is not effectively exchanged for monovalent hydrogen ions held by *Sphagnum* cell walls (or humic colloids), and it is easily displaced by cations of higher valence, such as calcium. Potassium is sparsely represented in fens because of such preferential cation exchange. Its availability is actually greater in bogs than in fens, even though it is scarce in ombrotrophic peatlands because precipitation water is potassium poor. Potassium can be taken up in some quantity when large amounts of potash (potassium carbonate) become available after a burn. The potash quickly washes away, but some quantities are carried by runoff into boggy depressions. Potassium, like phosphorus, is accumulated in cells by active uptake. In Finnish ombrotrophic mires, it has been demonstrated that the moss layer takes up and accumulates more potassium than phosphorus or nitrogen (177, 180).

Sulfur

Sulfur is held in plant cells as inorganic sulfates and organic combinations, including some amino acids. It is inactive at ordinary temperatures, and yet sulfur compounds occur plentifully in air, water, and soil, and the nutrient requirements of plants are almost always met by an abundance of sulfates. The generous supply of sulfur compounds can be attributed to bacterial conversions and atmospheric loads of sulfur dioxide derived in part from the burning of fossil fuels. Organically held sulfur is eventually freed by decay. Under aerobic conditions this oxidative process yields carbon dioxide and water in addition to nutritive ions:

$$\text{"CHONS"} + O_2 \rightarrow CO_2 + H_2O + NH_4^+ + SO_4^{--} + \text{energy}$$

Under anaerobic conditions hydrogen sulfide is produced, together with carbon dioxide, methane, and ammonium:

$$\text{"CHONS"} + H_2O \rightarrow CO_2 + CH_4 + NH_4^+ + H_2S + \text{energy}$$

Hydrogen sulfide readily takes up oxygen and is therefore toxic to aerobic organisms. If ferrous iron is present, hydrogen sulfide reacts with it to form the insoluble ferrous sulfide (FeS), which bars lake bottom sediments lower down from oxidation-reduction reactions. (Other ferrous, or reduced, compounds are soluble, except when sealed off by the sulfide.)

Sulfur-reducing bacteria get oxygen for the combustion of organic matter by reducing sulfate to sulfide.

Sulfur-oxidizing bacteria are of two general kinds:

1. *Colorless chemosynthetic aerobes* oxidize hydrogen sulfide to form sulfur and water and then oxidize sulfur to sulfate:

$$2H_2S + O_2 \rightarrow 2S + 2H_2O + \text{energy}$$
$$2S + 3O_2 + 2H_2O \rightarrow 2H_2SO_4 + \text{energy}$$

Sulfide minerals, such as ferrous sulfide, can also be oxidized by certain bacteria to elemental sulfur:

$$2FeS + O_2 + 2H_2O \rightarrow 2Fe(OH)_2 + 2S + \text{energy}$$

The energy of oxidation is used in the elaboration of carbohydrates.

2. *Colored photosynthetic bacteria* are anaerobic. Green sulfur and purple sulfur bacteria use hydrogen sulfide in place of water and give off sulfur instead of oxygen.

$$6CO_2 + 12H_2S \rightarrow C_6H_{12}O_6 + 12S + 6H_2O$$

Purple nonsulfur bacteria get the hydrogen needed for photosynthesis from organic compounds. The various purple photosynthesizers use dim, even invisible light, in the far red and near infrared wave lengths.

Iron

Iron is needed only in trace amounts, and it may be present in peat soils in very small quantities. It is precipitated under conditions of aeration, but its availability varies with pH. If the pH is high, iron may be precipitated out of solution. It exists in an oxidized ferric state with a valence of 3 and a reduced ferrous state with a valence of 2. Under oxygen-poor conditions it occurs in ferrous compounds, most of which are soluble. Hence, in anaerobic peat and at the bottoms of thermal-stratified lakes, iron is soluble (except in the form of ferrous sulfide). In acid, oxygen-rich conditions it exists in the ferric, insoluble form. In aerated, though less acid conditions, it becomes less soluble and at pH 7.5–7.7 entirely immobile. In the shallow water of fens, at a pH of about 6, iron precipitates as an oily film of ferric hydroxide and subsurface orange-brown gels of ferric carbonate, phosphate, and humate. Iron deposits in the beds of former lakes of alkalinity are made up of these same compounds. (Such deposits, as well as ferric iron precipitated under acid conditions, are known as bog iron. The deposits in the New Jersey pine barrens were important sources of iron during the American past. More recently, during the Second World War, the Germans imported bog iron from Sweden.)

Iron (like manganese, copper, and zinc) can be complexed, or chelated, with humic and tannic acids. Complexing provides for a slow and continuous release of nutrients. (The intense yellow-brown color of tannin-rich, dystrophic bog water is, in part, associated with complexed iron.)

Impermeable iron pans cause the waterlogging of slopes in areas where the climate is conducive to the formation of blanket bogs. The heath plants covering such slopes, together with *Sphagnum*, produce an acid litter that decomposes and causes the reduction of sesquioxides of iron and aluminum (changing trivalents to bivalents) and the formation of metal-organic complexes of some solubility. Fine clay, humic materials, and metal oxides wash down from the surface soil and form a hardpan in the subsoil. Clay and humic materials hold iron and aluminum by adsorption and chelation and keep them from further leaching. Red pans are indicative of ferric oxide and a degree of aeration. Sticky blue-gray deposits indicate ferrous oxides resulting from oxygen deficiency. In soils with changing water tables, pans form as alternating bands of red and blue.

Both bacteria and green plants cause carbon dioxide–oxygen changes that alter solubility. The products of bacterial action, such as sulfuric acid, also promote solubility. The soluble sulfates of aluminum and ferrous iron are among the worst of the toxic components of bog waters.

Iron often occurs in peat, in sulfide, sulfite, and sulfate combinations, and it is sometimes involved in the chemosynthetic elaboration of carbohydrates. Chemosynthetic bacteria get energy by oxidation-reduction reactions to build organic matter from carbon dioxide or dissolved bicarbonates. They make use of such starting substances as ammonia, sodium nitrate, sulfur, hydrogen sulfide, and ferrous iron combinations. Some examples are

$$2S + 2H_2O + O_2 \rightarrow 2H_2SO_4 + \text{energy}$$
$$2H_2S + O_2 \rightarrow 2S + 2H_2O + \text{energy}$$
$$2FeS + O_2 + 2H_2O \rightarrow 2Fe(OH)_2 + 2S + \text{energy}$$
$$4Fe(HCO_3)_2 + O_2 + 6H_2O \rightarrow 4Fe(OH)_3 + 4H_2CO_3 + 4CO_2 + \text{energy}$$

Chemosynthetic iron bacteria are generally restricted to wet, oxygen-poor habitats.

Ferrous sulfide sometimes forms as a whitish to greenish coating on plant surfaces in rich fens subject to a seasonal flooding. The precipitate forms in response to evaporation and its effect on pH.

Calcium

All plants need calcium, but they differ widely in their requirements and tolerances for it. The predominant inorganic compound in most freshwater systems is calcium carbonate, $CaCO_3$. It is also one of the least soluble. However, in the presence of carbonic acid it is readily altered to soluble calcium bicarbonate. Carbonic acid, H_2CO_3, is formed when carbon dioxide is dissolved in water. A gain or loss of carbon dioxide, because of temperature changes or biological events, shifts the carbonate-bicarbonate equilibrium. Marl may be deposited by the photosynthetic action of aquatic plants, including algae, removing carbon dioxide from calcium bicarbonate and precipitating calcium carbonate, or it may result from the loss of carbon dioxide to the air because of warm weather or abrupt changes in temperature, by seepage of cold, limy water into a warm lake, for example. In rich fens, marl is deposited as "pond porridge" in shallow pools subject to drying.

Rainwater would have a pH of about 5.6 if only carbonic acid were involved, but industrial pollutants, including especially the oxides of sulfur and nitrogen, acidify the rain and snow much further. The recent increase in acidity, owing to atmospheric pollution, has had a drastic effect on ecosystems developed on acid rock, but where soils have a calcareous component, lake waters have taken on greater alkalinity (105, 127). Acid rain and snow significantly increase calcium bicarbonate drainage into lakes and streams. The pH is controlled by bicarbonate buffering resulting from the rapid dissociation of carbonic acid into carbonate and bicarbonate ions:

$$H_2CO_3 \leftrightarrow H^+ + HCO_3^-$$
$$HCO_3^- \leftrightarrow H^+ + CO_3^{--}$$

After equilibrium is established, both the carbonate and bicarbonate ions dissociate:

$$HCO_3^- + H_2O \leftrightarrow H_2CO_3 + OH^-$$
$$CO_3^{--} + H_2O \leftrightarrow HCO_3^- + OH^-$$

The addition of more hydrogen ions, even from weak carbonic acid, neutralizes the hydroxyl ions formed by carbonate and bicarbonate dissociations. Hydroxyl ions may react with bicarbonate ions to form carbonate ions that combine with calcium and precipitate as marl:

$$HCO_3^- + OH^- \rightarrow CO_3^{--} + H_2O$$
$$Ca^{++} + CO_3^{--} \rightarrow CaCO_3$$

The pH remains unaltered as long as carbonate and bicarbonate ions are in equilibrium. Mineral-rich lakes and fen mats adjacent to them are kept circumneutral by carbonate-bicarbonate buffering, but bogs get only cation-poor water from the atmosphere and are low in calcium carbonate or bicarbonate. As a result, they are poorly buffered and strongly acid.

As the calcium concentration and the pH increase, microbial

activity and the formation of nitrates and sulfates increase. Nitrifying and nitrogen-fixing bacteria are stimulated by elevated levels of calcium. It follows, therefore, that the nitrogen available for plant use is greater in fens than in bogs. Calcium content is useful in separating peatland types as it correlates well with pH and decreases from fen to bog conditions. Groundwater concentrations of calcium may equal 20–30 ppm or more in fens, whereas bog water holds a mere 0.3–2 ppm. Water at the circumneutral level, as in fens, contains not only more, but more available nutrients, including calcium, than the water of acid bogs. But if the pH is too high, because of excessive lime, for example, iron may be precipitated out of solution, and certain other minerals, such as copper, manganese, and zinc, may also be rendered insoluble. On the other hand, lime increases the solubility of phosphorus and molybdenum.

Because calcium is needed for cell wall formation, as a component of the middle lamella, a calcium deficiency has considerable significance in the growth of root tips and hence root function. Although the concentrations of calcium (and magnesium) in bogs are low as compared with fens, leaf assays of ericaceous bog plants show more than adequate amounts of calcium, and the calcium content of bog peat is surprisingly high even at pH readings of 4. Since peat is highly adsorptive, much of the calcium is held in an exchangeable form available to roots of plants. Under deficiency conditions, some nutrients, like potassium, move from older leaves to growing points, but calcium does not.

Sphagnum requires a relatively low concentration of calcium, and in fact most species are unable to tolerate the combined effect of high acidity and high calcium. Species of hummocks are particularly sensitive to calcium, but those of bog hollows and swamp depressions are more tolerant. *Sphagnum teres* and *S. subsecundum* on sedge mats at the margins of calcium-rich lakes are protected from submersion and calcium excess by the fact that the mat floats up or down as lake levels change.

Though sufficient for the plants of poor fens and bogs, calcium and magnesium are not adequate at pH 5.5 or lower for a vegetation that will supply herbivorous vertebrates with the amounts required for skeletal structures. Rich and intermediate fens provide good forage, but the poorer the peatland is in nutrient ions, the poorer the fauna.

Magnesium

As a component of chlorophyll, magnesium has obvious importance to plants. Fortunately, it is not likely to limit plant growth in bogs and fens. A high proportion of the element occurs in salts or in an exchangeable form adsorbed to colloidal humic substances. In this way magnesium and calcium are similar, but magnesium compounds are generally more soluble than their calcium counterparts. Because of a greater abundance of calcium in sedimentary rocks, fresh waters usually receive more calcium than magnesium. But as calcium carbonate precipitates, owing to changes in the carbon dioxide balance, the ratio of magnesium to calcium increases. Like many other minerals, magnesium is more available in acid soils, although soils of alkalinity may have a much greater total content.

Carbon

Atmospheric carbon dioxide fixed by photosynthesis enters, secondarily, into the many carbon-containing compounds that living beings need. It is returned to the atmosphere by the respiration of plants, animals, and organisms of decay and, to an ever-increasing degree, by the combustion of fossil fuels. The use of coal and oil has increased significantly since the middle of the last century and especially during the past two or three decades. A vast amount of carbon is stored in peat reserves throughout the world. Carbon dioxide locked up in oxygen-starved peat is not likely to be freed by decomposition as long as cool, wet, acid conditions prevail. But should the climate become warmer and dryer, as it has in the postglacial past, the decay of peat would be greatly accelerated. The world would become a hothouse covered over by a carbon dioxide ceiling allowing sunlight to penetrate but preventing heat loss by radiation.

The carbon dioxide content of the atmosphere has been increased by industrialization dating from about 1870, but much of it is too young, judging from radioactive decay, to have come from that source alone. It is possible that a warming trend over the last century has already set the world's peat to a slow burn by encouraging oxidative decay. At the present rate of accumulation, the carbon dioxide content of the atmosphere might create a warmth sufficient, perhaps within the next four hundred years, to cause all the world's glacial ice to melt. Meltwater and thermal expansion of water may cause oceans to rise as much as 200 feet. Appreciable changes in ocean levels within the next twenty to thirty years may give warning to New York, London, Tokyo, and other coastal cities worldwide to move upland.

Oxygen

Anaerobic conditions prevail at or somewhat above the water table, which in bogs stands about 8 inches below the surface. Oxygen deficiency slows down ion uptake and metabolism. However, plants of wet habitats are likely to be provided with intercellular airspaces and spongy tissues. Dissolved oxygen concentrations may be very low and still sufficient to maintain oxidizing conditions, yet there is a level at which nutrient cycling owing to decay is limited by a shortage of oxygen. Where shortages result from waterlogging, oxygen gradients can be demonstrated from leaves to petioles to stems to roots. The carbon dioxide gradient is just the opposite. The problems of carrying out respiration are magnified by the slow movement of dissolved oxygen through peat. The oxidation-reduction balance varies with fluctuating water levels.

Some plants, for whatever reason, cope with oxygen deficiency better than others. In British peatlands, for example, *Eriophorum* stands waterlogging better than *Molinia*, *Erica tetralix* better than *E. cinerea*. Vegetational zonations in peatlands are evidence of differing tolerances to water-soaking and oxygen shortage. Plants of wet fens may have roots better suited to oxygen shortages than bog species. A buildup in carbon dioxide triggers a reduction in oxygen use, lowering respiration rates before anaerobic conditions cause toxins to accumulate. Reducing conditions can cause glycolysis to be accelerated and ethanol to be produced, but in flood-tolerant plants, glycolysis is not hastened and nontoxic malate is produced instead of ethanol. Aeration brought on by periodic changes in water level or internal flow may reduce a toxic buildup of ethanol, and oxygen leaks from roots and rhizomes may have a similar effect.

The Frugal Use of Nutrients

Perennial herbs lose nutrients by the annual decay of leaves, yet they also provide for spring renewal by moving foodstuffs and nutrients to winter buds or underground storage organs. Some wetland plants—*Eriophorum vaginatum, Rubus chamaemorus, Typha,* and no doubt many more—move nutrients to stems and roots before the onset of winter dormancy. With a new season's growth, nutrients are rerouted to the meristems, young leaves, and eventually flowers and fruits. The evergreen ericads that predominate in bogs very likely save mineral resources by similar translocations. Birch, alder, and tamarack, however, lose their leaves each year and do not save nutrients by translocation. Leaf fall returns nutrients to the peatland each year when plant growth ends and nutrients are not needed. The nutrients held in leaf litter are recycled by decay, but they are also made subject to loss by leaching and flooding. Evergreens, by contrast, lose leaves gradually and thus guard their resources from sudden decay and total loss by seasonal water flux.

Carnivory

Insectivorous plants use animal proteins as a dietary supplement. They often grow in acid habitats where bacterial recycling is minimal and where root systems have little access to mineral nutrients. (Some of them, however, have a broad tolerance and are even more commonly represented in more mineral-rich peatlands.) Pitcher plants, of the genus *Sarracenia* (figs. 4-2, 4-10), have long, hollow leaves partially filled with water. They are death traps lined with stiff, downward-directed hairs that prevent insect escape. Species of *Drosera* (figs. 4-3–5, 4-7), aptly called sundews, have leaf blades covered with hairs that secrete glistening drops of stickiness. Insects are trapped, flypaper fashion. *Pinguicula* (figs. 4-6, 4-11), called butterwort because its leaves can supposedly sour milk, has leaves covered by stalked glands that secrete a stickiness and sessile ones that provide digestive enzymes. The bladderworts, species of *Utricularia* (figs. 4-8–9), have underwater leaves hollowed out and fitted with a trapdoor. Sensitive hairs trigger a sudden expansion and opening of the bladder leaves, causing the prey to be sucked in. *Dionaea* (fig. 4-12), Venus's flytrap, a rarity of the coastal southeastern United States, has leaves that close up like a steel trap.

The species of *Sarracenia* attract insects by the bright colors of their pitcherlike leaves, as well as nectar secretions, especially at the mouth of the pitchers. Slippery cuticles, stiff, downward-directed hairs, and a numbing secretion contribute effectiveness to the trap.

Slow digestion in the pitcher solution results from weak secretions of enzymes but more so from bacterial activity.

At least seventeen arthropods are obligate associates of *Sarracenia* (215), able to inhabit the leaves yet escape digestion. Arthropods that feed on microorganisms and particulate matter in suspension in *S. purpurea* leaves include a mite and larval stages of a mosquito, a midge, and a sarcophagid fly. (There is reason to think that the invertebrates aid in the host's digestive processes, although bacterial exoenzymes are also involved.) Insects feeding on leaf tissues include an aphid and two moths.

It is curious that *Sarracenia* and other carnivores especially flourish in mineral-rich peatland sites where nutrients would seem to be in ample supply. But the availability of phosphorus in minerotrophic niches where iron and aluminum may render it insoluble may make carnivory a useful strategy in fens as well as in mineral-poor bogs.

Parasitism

The dwarf mistletoe, *Arceuthobium pusillum* (figs. 4-13–15), solves its nutritional problems by parasitizing black spruce and, less commonly, white spruce. (It also occurs, rarely, on tamarack and red pine.) It is seen on young twigs as stubby, brown outgrowths scarcely resembling a flowering plant. Its presence can also be detected by dense branching of the host in the form of witches' brooms. The infection reduces the tree's growth, in height and diameter, and heavy infection causes a weakness in the tree, so that it is likely to be killed by drought, insects, or fungi. *Arceuthobium pusillum* causes serious damage to white spruce in Maine, and elsewhere.

Root/Fungus Mutualisms

Root/fungus associations, or *mycorrhizae*, occur in about 80 percent of all angiosperms, and similar dependencies occur among gymnosperms, pteridophytes, and even some of the rootless liverworts. Most land plants take advantage of the superior ability of fungi to absorb minerals from the soil. Mycorrhizae have a limited significance in wet, acid peatlands, but the Ericaceae and Orchidaceae are mycorrhizal, and mycorrhizae have been demonstrated in tamarack, black spruce, and other woody plants of bogs and fens. (It is curious that one family characteristic of peatlands, the Cyperaceae, have no such associations.)

Acid peatlands are totally unsuited to all but a few plants, yet those few appear to live there comfortably and even thrive. Freedom from competition is a factor in their success, and so are certain physical and physiological adaptations. Increased absorption of nutrients owing to a fungal root symbiont can be counted as one such adaptation. Very importantly, mycorrhizae increase the uptake of nutrients most often in short supply—nitrogen, phosphorus, and potassium. The fungus gives over nutrients in addition to cytokinins, gibberellins, and vitamins; and it takes back simple sugars. The fungal hyphae take over the function of root hairs (which fail to develop). Some of the hyphae penetrate cortical tissues and trigger an increase in the amount of carbohydrates moving to the roots. Though not absolutely dependent on the fungus, the "host" benefits in mineral status and growth. Some mycorrhizal fungi also protect the host from soil toxins. Some and perhaps all mycorrhizal fungi, on the other hand, are wholly dependent on their hosts for carbohydrates. The result of this symbiosis is that both host and fungus can exploit habitats unsuited to either alone. Mycorrhizae characterize poor soils. Both the degree of infection and the benefits exchanged are greater in nutrient-poor soils.

Only a few mycorrhizae have been studied in detail, and the fungal symbionts are, for the most part, unknown except by circumstantial evidence. There are three general types of mycorrhizae.

1. *Ectomycorrhizae* (fig. 4-16) occur in only about 3 percent of all plant species, yet they have considerable significance because the majority of them involve forest trees of temperate latitudes. The fungus surrounds young roots as a bulky mantle. The mantle is efficient in absorbing and accumulating nutrients, and the absorptive surface is sometimes increased by hyphae ramifying into the soil. The short lateral roots enveloped by a mantle take on a stubby appearance. If infected at all, the main roots remain free at their growing tips. Septate hyphae (of basidiomycetes and ascomycetes) penetrate the root and grow among the cortical cells as a haustorial net.

The genera of fungi known to be involved in ectomycorrhizae include ninety-one basidiomycetes, mostly agarics, seven ascomycetes, and two Fungi Imperfecti. Such familiar basidiomycete genera as *Amanita*, *Boletus*, and *Russula* are almost exclusively mycorrhizal. Host genera include thirty-nine angiosperms, ten gymnosperms, and two ferns. Peatland plants known to have ectomycorrhizae include *Larix laricina*, *Picea mariana*, *Pinus banksiana*, *Alnus incana*, *Betula glandulosa*, and *B. pumila*. Most of numerous fungi presumed to form mycorrhizae with those plants occur in Michigan's peatlands but are not restricted to them.

Ectomycorrhizal fungi are, generally speaking, incapable of living separate from their hosts or decomposing organic matter rapidly, if at all. They are outcompeted by other fungi for simple sugars in the soil. They can take up amino acids and thereby take advantage of proteins broken down by other microorganisms. They have some capacity, in culture, for hydrolyzing organic phosphates.

2. *Endomycorrhizae* (figs. 4-17–19) are by far the most common of the root/fungus associations. The majority have nonseptate hyphae of the phycomycete family Endogonaceae, but the fungi associated with the Ericaceae and some of the Orchidaceae have septate ascomycete and basidiomycete hyphae. The fungal hyphae are not much aggregated around the roots. They grow among the cortical cells of the roots and also penetrate them. The hyphal tips may be coiled, bushy-branched, or swollen and oil-rich. Starch disappears from root cells with which these structures are associated, and nutrient transfer results from the digestion of the hyphal structures.

Presumably all the Ericaceae are mycorrhizal, and the fungi associated with them are, for the most part, cup-forming ascomycetes. Such peat-growing genera as *Calluna*, *Erica*, *Rhododendron*, and *Vaccinium* (including the cranberries, at least) have this kind of mycorrhizae. The roots terminate in fine, unsuberized rootlets that are commonly moniliform-constricted and have only one to three layers of cortical cells. The mycorrhizae increase the host plant's nitrogen and phosphorus content and improve seedling growth. The fungi can use organic phosphorus in culture and therefore appear to have some decomposer role.

Orchids are never free of mycorrhizae. Some have phycomycete hyphae, lacking septa, and others have septate basidiomycete hyphae. The fungi have an extensive saprobic existence, and it appears that the orchids simply parasitize them. The dustlike seeds, virtually lacking an endosperm, depend on a fungal source of carbohydrate for germination and seedling growth. In nongreen orchids, such as *Corallorhiza*, the fungus serves as a bridge between the orchid and a carbohydrate-giving tree with which the fungus has a mutualistic relationship.

3. The intermediate *ectendomycorrhizae* form a mantle as in ectomycorrhizae and penetrate the cortical cells as in endomycorrhizae. They occur in some of the Ericaceae, *Arbutus* and *Arctostaphylos*, for example, and in the related Pyrolaceae and Monotropaceae. A basidiomycete (*Boletus*) provides a nutrient bridge from the roots of pines and spruces to the nongreen *Monotropa* (fig. 4-20).

Peat Mosses and Fungi

Sphagnum spores germinate readily in culture, in the absence of fungi, yet the germination depends on phosphorus, and so it seems reasonable to speculate on a fungal source of phosphorus in a habitat notoriously low in that nutrient. Both *Sphagnum* and *Sphagnum* peat are said to contain substances that inhibit the growth of roots and the germination of *Sphagnum* spores. These substances are destroyed by mycorrhizal fungi associated with the Ericaceae and also nonsymbiotic fungi regularly present in bogs, such as *Penicillium*, *Verticillium*, and *Mortierella* (27).

Sphagnum magellanicum is much involved in the fen-to-bog changes associated with hummock formation. It is said that fungi tend to infect and degrade *S. magellanicum* at pH values greater than 5 (Cyrus McQueen, personal communication), and this may explain why that species does not invade sedge mats but assumes prominence in the more acid stages of succession.

In summation, the communities that occupy peatlands depend, absolutely, on the chemical environment and the possibility for nutrients to be made available. Peat is a nutrient lockup. As peat accumulates, so do the nutrients on which life depends. Nutrient cycling depends on decay, and as decay is limited, so are the available supplies of phosphorus, nitrogen, and potassium and so are the kinds of plants that make do with shortages, that eke out a living by cation exchange, active uptake, internal translocation, symbiotic interdependence, carnivory, and parasitism. The effective use of atmospheric input, scant as it is, makes possible the continued development from fen to bog.

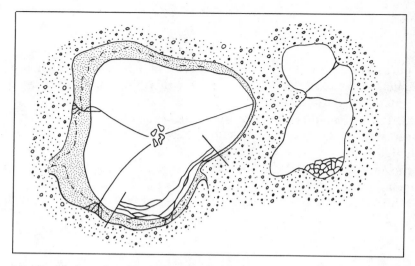

Fig. 4-1. Animal tracks channel nutrients from the lagg into oligotrophic bog regions. (Redrawn from Worley 1981.)

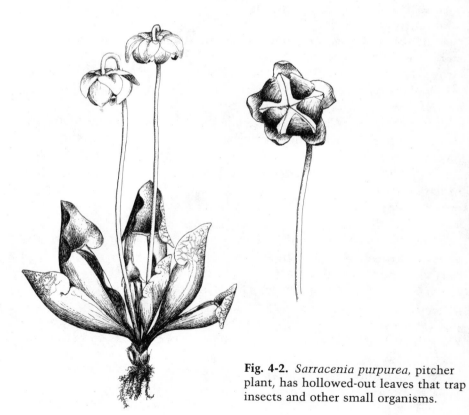

Fig. 4-2. *Sarracenia purpurea*, pitcher plant, has hollowed-out leaves that trap insects and other small organisms.

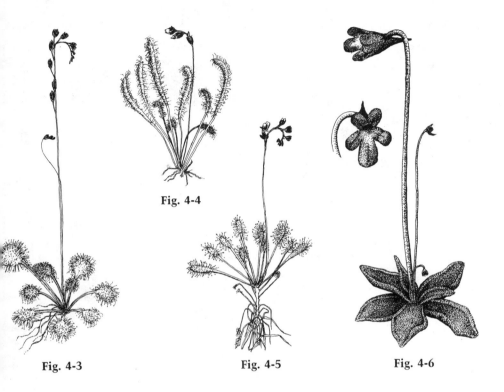

Figs. 4-3–5. Sundews. **Fig. 4-3.** *Drosera rotundifolia.* **Fig. 4-4.** *Drosera linearis.*
Fig. 4-5. *Drosera intermedia* (showing the characteristic long stem).
Fig. 4-6. *Pinguicula vulgaris,* butterwort. The flowers are purple and the leaves yellowish.

Fig. 4-7. A single leaf blade of *Drosera intermedia*, one of the sundews, showing sticky glands by which insects are trapped

Fig. 4-8. The green leafy floating portion of a species of *Utricularia*. (Photo by Kerry Walter.)

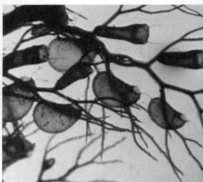

Fig. 4-9. The submerged portion of *Utricularia intermedia* showing leaves modified as bladders that trap small arthropods. (Photo by Kerry Walter.)

Fig. 4-11. *Pinguicula vulgaris*, butterwort, in flower

4-10. Leaf of the cher plant, *Sarracenia rpurea*

Fig. 4-12. *Dionaea muscipula*, Venus's flytrap, endemic to peatlands of the Carolina coast, has a steel-trap mechanism for catching insects.

Figs. 4-13–14. *Arceuthobium pusillum*, dwarf mistletoe, parasitic on black spruce. The flowers are insect-pollinated e in May.

Fig. 4-13. Female

Fig. 4-14. Male

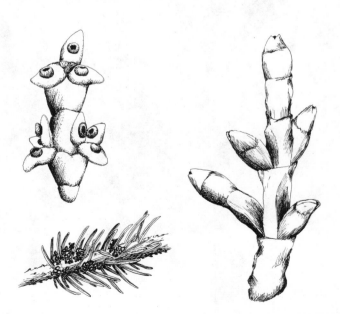

Fig. 4-15. *A. pusillum*, dwarf mistletoe, on a young branch of black spruce, male plant (upper left), female plant (right)

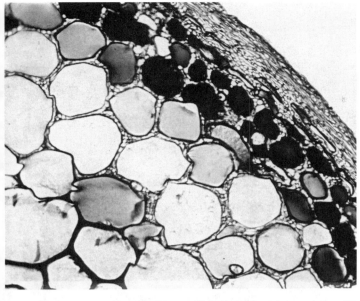

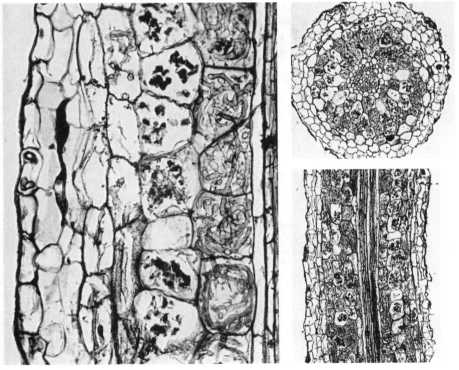

Fig. 4-16. A young root of *Tsuga canadensis* in cross-section, showing the mantle and intercellular hyphae of an ectotrophic mycorrhiza. (Photo by E. B. Mains.) **Figs. 4-17–19.** Cross- and longitudinal sections of a *Liriodendron* root, showing fungal penetration of cortical cells in an endotrophic mycorrhiza. This kind of mycorrhiza is common among peatland plants. (Photos by E. B. Mains.)

Fig. 4-20. *Monotropa uniflora*, Indian pipe, a species of a variety of woodland habitats including spruce muskegs, depends on a 3-way mycorrhizal relationship, *Monotropa*-fungus-tree.

Chapter 5

Life in a Wet Desert

> Cold and damp—are they not as rich an experience as warmth and dryness?
> —Thoreau

If plants are able to grow in water or in wetlands at all, it is because of unusual adaptations, physical and physiological, for life under stress. With too much water go problems in aeration and mineral nutrition. The plants that grow in fens and bogs cope with varying degrees of environmental rigor, and for that reason they distribute themselves according to gradients in moisture, nutrient availability, and pH. A considerable problem in survival has to do with seasonal and diurnal temperature extremes that affect water balance. Pioneer plants of peatlands are selected by the physical and chemical conditions of the environment, but as fens become bogs the plants determine the composition and transformation of the communities characterizing each stage of succession.

The Lake Environment

> A lake is the landscape's most beautiful and expressive feature. It is earth's eye; looking into which the beholder measures the depth of his own nature.
> —Thoreau

Because of their small size and location in depressions, the lakes where mire systems develop are commonly protected from high winds, ice floe damage, and seasonal upheavals when cold surface waters plunge downward causing oxygen and nutrients to be redistributed. Relative quietness allows sediments to accumulate and marginal vegetation to encroach on them. The record of vegetational change, in fact, the whole history of the fill-in process, is one of lake senescence. The entire history is recorded in the sediments.

Lake-bottom sediments are fine-grained because of size-sorted inorganic materials transported from littoral zones and because of a muddy organic detritus. Mineral components are derived from the substrate and from the silicon of diatom shells and lime precipitated, as marl, by biologically induced shifts in the carbonate-bicarbonate equilibrium. In eutrophic lakes oxygen deficiencies develop, and dense populations of algae and invertebrate animals feeding on algae contribute by death and sedimentation additional muddiness. The humic acids associated with incomplete decay eventually combine with calcium and magnesium to add an amorphous organic content to the bottom ooze, and bottom faunal deposits include fecal nitrogen and phosphorus. The bottom mud, known as gyttja, is gray to dark brown. In deep water it becomes compact and gelatinous, and under conditions of anoxia it is made shiny-black by ferrous sulfide and smelly by hydrogen sulfide. The inner layers of mud lock nutrients away, but the mud surface can release them under conditions of seasonal aeration.

Most iron precipitated to the depths of thermal-stratified lakes is retained there, but the bottom muds allow some iron to change from reduced to oxidized states in response to oxygen circulation during spring and fall overturns. The shifts in solubility associated with aeration, in compounds of iron, manganese, and other metals, make some anions, such as phosphate, also available. The spring overturn brings soluble phosphates to the surface in concentrations that promote algal blooms and at the same time brings oxygen to the bottom muds where phosphorus is rendered insoluble. With a return to conditions of anoxia, phosphorus at the lake depths is again solubilized.

Humic acids chelate iron and keep it available for slow release under pH and aeration conditions in which it would normally precipitate (as ferric iron compounds). They also modify anion availability. The precipitation of phosphates entering into combination with iron, aluminum, calcium, and other cations may be delayed by the cation-exchange capacity of humic colloids. As an example, when calcium is the dominant ion adsorbed by such colloids, phosphorus can remain available in solution.

The formation of lake sediments (fig. 5-1) largely depends on productivity, and that, in turn, on the supply of nitrogen and phosphorus. All other elements are commonly present in excess of biotic needs. The nitrogen cycle is relatively complex because nitrogen exists in a free state and also in nitrite, nitrate, and ammonium

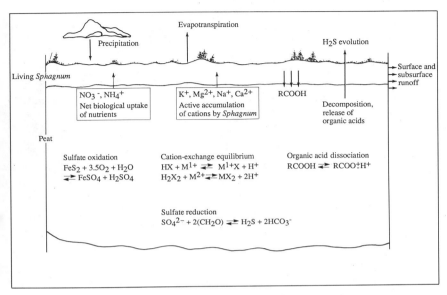

Fig. 5-1. The chemical transformations that take place in a bog lake. It appears that the lake gains in acidity not because of strong acids derived from atmospheric pollution or the ion-exchange capacity as much as the dissociation of organic acids (RCOOH) resulting from the slow decomposition of anaerobic peat. (Redrawn from Kilham 1982a.)

combinations, not to mention organic combinations, and some bacteria derive energy by oxidizing one nitrogenous compound to another. The phosphorus cycle is simpler. Phosphorus occurs in only one oxidation state, as phosphate. During the summer the plant and animal populations in lake water are at their peak, and dissolved phosphate is used up. Also, the algae are eaten by zooplankton whose fecal pellets settle out and remove phosphorus from the upper water. Some phosphorus is cycled back into use by planktonic death and decay, but the cold lower water does not mix with the warm upper water where photosynthesis takes place except at spring and fall overturns. The summer shortage of oxygen at the lake depths causes phosphorus to become soluble because of the reduced ferrous iron (Fe^{++}). On overturn, soluble phosphate is carried upward, and oxygen brought downward causes iron to change to the ferric state (Fe^{+++}), which renders phosphate insoluble. The cycling of phosphorus with oxygen circulation renews growth and productivity.

A high productivity linked to mineral-rich waters causes an abundance of sediments and a quick, centripetal fill-in. Lakes broad enough for a fetch of wind may have waves that distribute sediments to an even overall depth inconducive to mat formation at the margins.

Eutrophic bog lakes often hold a suspension of incompletely decomposed fragments of sedge peat. These and other bits of organic matter eventually form a false bottom that may be many meters deep and in especially dry years exposed near the lake margins or much more extensively as a mud of little substance, but enough to allow a pioneering advance of sedges (figs. 2-10–12). Sediments accumulate as productivity increases, especially because decay is slowed down as oxygen is depleted by increased populations of aquatic organisms.

Trophic levels in mineral-poor and mineral-rich lakes are indicated by numbers and kinds of plankton (207). Oligotrophic lakes are relatively poor in algae, and the overabundance of individuals associated with water blooms is rare. The species are, however, fairly numerous and distributed to great depths. Characteristic algae include *Staurastrum* and (in calcium-deficient water) many desmids, as well as the diatoms *Tabellaria* and *Cyclotella* and the chrysophycean *Dinobryon*. By contrast, eutrophic lakes have a greater mass of algae, and water blooms are frequent. The species are numerous, but some are poorly represented and easily overlooked. They are concentrated at aerated upper levels. Characteristic are the blue-greens *Anabaena* and *Aphanizomenon* and also the diatoms *Fragilaria*, *Stephanodiscus*, and *Asterionella*. There are many indicators of eutrophy, but few that are indeed limited to conditions of oligotrophy. The latter are species of broad tolerance that are able to survive nutrient-poor conditions.

Humic compounds contribute most of the darkness to peatland water. In northern lakes, especially those of acid substrates deficient in calcium, on the Canadian Shield, for example, the water may be stained yellow-brown and reduced in transparency because of dissolved and suspended organic matter, especially tannic acids leached from nearby peat-forming vegetation. Such lakes, known as *dystrophic*, may be highly acidic, at the farthest extreme from the eutrophic and highly productive lakes with calcium bicarbonate buffering. However, the term also denotes lakes formed on more nutrient-rich glacial outwash where the water is stained by tannins leached from white cedar and leatherleaf vegetation (and not necessarily bog mat vegetation). Tannins are secondary plant products, soluble in water. They remove ions from solution as humic acids do, by complexing. That, rather than acidity, explains why dystrophic lakes are generally unproductive. (A dystrophic lake, as a matter of classification, is one that is entirely outside the eutrophic-oligotrophic series of phosphorus availability.)

Temperatures, High and Low

The bog ericads—*Chamaedaphne, Andromeda, Kalmia, Vaccinium,* and *Ledum*—cope with an environment of extremes. These same genera range beyond the bog habitats of the upper Great Lakes to the heaths and muskegs of the north where climates are even more harsh and forbidding. Some of the *Vaccinia* of bogs lose their leaves and overwinter in obvious dormancy, but most bog shrubs display other features that seem to give protection from driving winds, abrasive snow, short growing seasons, and loss of water and nutrients. Most members of the Ericaceae deal well with acid, mineral-poor substrates, whether in habitats of water stress or winter extremes or not. A low and decumbent habit of growth and tough, leathery leaves seem suited to the blustery cold of winter. Devices to protect against water loss—thick, shiny, evergreen leaves with heavy cuticles, recurved margins, woolly undersurfaces, and sunken stomata—seem anomalous in a too wet bog environment. These apparent water savers may be better regarded as taxonomic characters of the family than strategies for survival. Investigators have failed to find decreased water loss in bog plants of xeromorphic character or demonstrate physiological drought (76, 131). Yet, water stress can be a real hazard to bog plants, not in the fine days of summer but in times of drought and in brief periods in winter when exposed leaves break dormancy or in spring when vigorous growth begins even though roots are still encased in ice. The water stress at such times of critical need is not easily evaluated in the laboratory.

In southern Michigan and Wisconsin, sedge meadows may suffer water stress during summer heat. Yet, they may be somewhat protected by a relatively low rate of evapotranspiration. Water lost by evaporation may be no more than one-fourth that lost in upland pastures exposed to warm, dry winds. Cold air drainage and heat loss by radiation during clear nights lower the temperature of lowlands

Fig. 5-2. Stutsmanville Bog, Emmet County, Michigan, morning fog associated with cool air resulting from nighttime heat loss by radiation, as well as cold air drainage. (Photo by Jeffrey Holcombe.)

(fig. 5-2), and some shelter from winds may continue the low night temperatures into the day. The reduced temperatures are responsible for a short growing season. Killing frosts may occur early in September in sedge meadows in the lower Great Lakes region, but they are much delayed in the surrounding uplands, even to the end of October. Northward, where the precipitation-evaporation balance is more favorable to bog development, differences as much as 20–30°F may exist between bogs and nearby uplands. In northern Wisconsin roughly seventy days are frost-free (48), except in boggy lowlands where frost is possible any night of the year. Because of the danger of summer frost, peatland weather forecasts are regularly broadcast so that Wisconsin cranberry growers can protect the ripening berries by flooding or spraying. The same danger of early frost occurs, of course, in other parts of the Upper Midwest.

The diurnal fluctuations in surface temperatures can be considerable owing to heat absorption by day and radiation by night. A summer reading at 3 feet above the bog surface may be about 80°F, at the surface 90°, and 3 feet below only 40° (152). Water temperatures in the flarks of northern Michigan's patterned peatlands in midsummer may be about 86°F, in contrast to 62–68° in peat 12 inches below the flark (140). (Wet peat is a much better insulator than dry peat, and *Sphagnum* takes on much less heat than dark strongly humified peat, or muck, does.) During the hottest days of summer, surface temperatures can reach 100°F at the same time that roots, no more than a foot below ground, are coping with temperatures of 30–50° or lower. Bogs remain frozen much later in the spring than the uplands do. Midday evaporation may counterbalance a degree of heat absorption by the dark-colored *Sphagnum* of bog mats. The surface topography of the bog and the growth form of bog shrubs, especially in hummock-hollow zones, are such that warm, moist air is trapped, thus offsetting some temperature extremes during the growth season. Compact cushions of *Sphagnum* heat up slowly, and even species of looser growth trap warm air and maintain temperatures slightly higher than those of the substrate and the surrounding air. Solar radiation and a lesser wind velocity at plant level account for daytime surface temperatures that are higher than air temperatures.

Deep snow often results from wind speeds modified by surrounding hills and forests. The snow pack (fig. 5-3) protects evergreen shrubs from exposure and provides them with temperatures higher than those of the air. The root levels may, in fact, be freezing while the leaves are functioning at springtime temperatures of 80–85°F or more (48). *Chamaedaphne, Andromeda,* and *Kalmia* flower in May, at a time when their roots may be frozen. In Maine bogs, ice has been observed at root level well into the summer, and in northern Minnesota scattered frost layers are seen in July, rarely in August.

The roots are insulated by wet peat. The heat capacity and conductivity of peat are correlated with water content. As any muck farmer knows, a dry soil is warm in the spring, but a wet one is cold. Draining muck, which is much decomposed reed-sedge peat, lowers its mass and therefore its specific heat, and less energy is then needed to raise the soil temperature. Evaporation and its cooling effect are also reduced by drainage. Owing to its blackness, muck absorbs heat quickly during the day, but cold air drainage and heat radiation cause nighttime temperatures to be low and early frost a certainty. Because of spring wetness and alternating freezing and thawing, muck is subject to frost heaving. The ability of roots to absorb water is much reduced at cold temperatures. Cold alters the viscosity of protoplasm, decreases membrane permeability, and retards root growth and root hair formation. Draining muck allows roots to penetrate more deeply and function at more favorable temperatures, and it also allows the soil to warm up quickly enough for use in areas otherwise unsuited for agriculture.

Water and Lots of It

Hydric successions are linked to a progressive decrease in wetness. It can be said, in fact, that the successive stages of vegetation cause that decrease. Relative humidities are greatest at ground level, and evaporation is progressively increased in layers of vegetation developed above the ground. As a plant grows upward, it encounters conditions of greater water stress. Shrubby heaths lacking a *Sphagnum* cover evaporate 40 percent less than those with *Sphagnum*. A tuft of *Sphagnum* evaporates five times as much as water of equal area (113), and the absorption by *Sphagnum* may be as great as twenty-seven times its dry weight. Not all species take up that much. The average may be less than twenty times. The considerable evaporation from the surface of an open *Sphagnum* mat lowers both the maximum and the minimum daytime temperatures, but the *Sphagnum* surface loses less water when sheltered by a layered vegetation. In fact, cushions and carpets of the same species of *Sphagnum* evaporate nearly 6.5 percent less when layered over by shrubs. Even small vascular plants, like cranberries, conserve moisture and protect *Sphagnum* from desiccation by creating humid air spaces only a few centimeters above the surface.

In the tundra, *Sphagnum recurvum* var. *tenue* shows a much higher rate of water loss than true mosses, lichens, and vascular plants. That same species and variety and other members of the section *Cuspidata* quickly show signs of water stress in peatlands elsewhere. The species of *Sphagnum* are positioned in hummocks and hollows according to their differing abilities to soak up and conduct water. The *Cuspidata*, submerged or loosely tufted in hollows, dry quickly and therefore seem to benefit by growing in wet or at least sheltered positions. The sections *Acutifolia* and *Sphagnum*

Life in a Wet Desert 149

Fig. 5-3. Brighton Bog, Livingston County, southeastern Michigan, in late December, is actually a poor fen in transition to a bog, supporting a scattered growth of *Larix* and a very few, very short *Picea mariana* trees in association with *Chamaedaphne*. The lake appears to be relatively acid, judging by the fact that the mat is pioneered by *Sphagnum* rather than sedges and rushes as most lake-fill peatlands in southern Michigan are. The insulating effect of a snow pack protects evergreen bog shrubs against water loss during the winter when the soil water is frozen.

hold water better and occupy the more exposed hummocks. *Sphagnum fuscum* at the dry tops of hummocks is well suited to a need to soak up water quickly. The dense-packed, wet *Sphagna* of hummock sides (*S. magellanicum* and *S. capillifolium*) are less suited to water movement, but they are good at water retention.

Wet *Sphagnum* moves water inefficiently, if at all, by capillarity. Hummocks of *Sphagnum* are limited to the height that *capillary water* can move against gravity, some 20 inches. At that height increased evapotranspiration also limits growth, and *Sphagnum* is replaced by more xeric mosses and lichens. The dark colors of hummocks, commonly brown at the top and red below, absorb heat. (The contrast in warmth between red or brown *Sphagna* and green or yellowish ones on a sunny day is obvious to the touch.) That heat causes an evapotranspiration pull on capillary water that aids in bringing minerals up to a height where they are in particularly scant supply.

Groundwater is derived almost entirely from rain and snowmelt soaking into the soil and slowly moving down to a zone where all pore spaces are filled. The top of this zone of saturation is the *water table*. Above it the pores are filled with air and capillary water films. The water table follows irregularities in surface topography because of frictional resistance to capillary water flow. In a bog, which has some convexity, however slight, the water table stands near the surface most of the season and follows that convexity (fig. 5-4), because of the impermeability of the underlying wet peat. The water table is unable to flatten out because of the continued addition of water from above.

Capillary movement depends on cohesion of water molecules and their adhesion to soil surfaces. It is favored by small pore sizes. Bog hummocks grow upward until capillarity is counterbalanced by gravity. In domed bogs the water table may be perched high above the regional water table, but capillary movement in relation to the water table is unaltered. Evapotranspiration is greater when the water table is some 4 inches below the surface, at a level favoring root function and constant capillary movement, than when the upper peat is saturated. But when the water table is drawn down below bog hollows, evapotranspiration is considerably reduced because capillary water cannot reach the root zone as much as 20 inches above the hollows. In particularly dry years, bog plants may suffer considerable water stress because of extreme drawdown of the water table (22).

Poorly decomposed *fibric peat* near the bog surface is coarsely porose, and water passes through it easily. Partly decomposed *hemic* and well-decomposed *sapric peats* are, by contrast, quite impermeable even though thoroughly soaked, and they hold water tenaciously. A hole dug in fibric peat fills with water within minutes, but a deeper one may take weeks to fill. All peat, fibric, hemic, or sapric, holds 80 percent or more by volume of water when saturated, but the rate of water flow and the amount of water retention are related to pore size and connection. More decomposed peats have smaller pores, move water more slowly, and retain more of it. The rate of water movement in sapric peat may, in fact, be slower than in

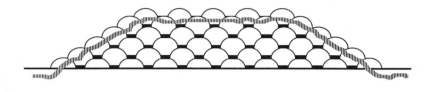

Fig. 5-4. A domed bog, showing the regeneration cycle whereby hummocks and hollows replace one another. The water table follows the bog surface and is therefore lifted, or perched, above the level of the regional groundwater.

some clays. The impermeability of sapric peat seals off the groundwater and causes precipitation water to perch above the regional water table and separate from it. Once drainage drops the water table to the level of sapric peat, ditching has little effect more than 15 feet away. But drainage through fibric peat is effective 150 feet away.

A sponge cannot take up water after it is wet. Likewise, once peat is wet, any added water runs off. Contrary to common belief, peat cannot soak up excess water or release it slowly and sustain stream flow during summer drought. Peatlands are, in fact, incapable of holding the major portion of the annual precipitation (22). Water from snowmelt and spring rains is largely lost by runoff, and most of the summer rain is lost by evapotranspiration. Except at times of rapid runoff, water exchange takes place above the summer level of the water table, in the zone of partial decomposition where the peat structure allows for capillary movement. Storage water brought up from the water table by capillarity is released by evapotranspiration. What storage capacity there is reduces the flow of water into streams during summer rains, but peatlands evaporate water at a maximum at the very times when drainage is most reduced.

The water that can be retained by a peat bog is enormous, but a point comes when no more can be held, when the weight of water added to a domed or blanket bog has to go somewhere. A bog burst can be as devastating as a flash flood. In County Kerry, Ireland, on the night of December 27, 1896, a bog of 600 acres, convex at the center and holding 5.5 million cubic feet of peat, suddenly gave way, sweeping away the steward's house and drowning him and his family. The deluge was attributed to heavy rain, as well as injudicious cutting of peat.

In contrast to a grounded mat, a sedge mat can adjust to fluctuating water levels. A Minnesota sedge mat on shallow peat can change its surface level as much as 2.3 feet per year, whereas a tamarack stand on deeper, firmer peat varies no more than 4 inches (26). The sedge mat at Inverness Mud Lake in Cheboygan County, Michigan, shows a similar annual fluctuation of 2.2 feet (80). The freedom to float allows a sedge mat to be in constant contact with water, yet not flooded.

Fen mats developed at the margins of lakes can receive upwelling water from underlying sand aquifers as well as from surface water and slope drainage. Such springs are easily detected in mild winters because their water does not freeze as readily as the rest of the fen surface. In more built-up peatlands, upwelling springs become plugged by peat accumulations.

Sphagnum Smothering and Mothering

> For just as the *Sphagna* suck up the atmospheric moisture and convey it to the earth, do they also contribute to it by pumping up to the surface of the tufts formed by them the standing water which was their cradle, diminish it by promoting evaporation, and finally also by their own detritus, and by that of the numerous other bog-plants to which they serve as a support, remove it entirely, and thus bring about their own destruction.
> —Schimper (as translated by Braithwaite)

Sphagnum takes the lead in the origin, development, and day-by-day biology of peatlands. It flourishes in acid, nutrient-poor habitats but can also live in circumneutral to alkaline environments where nutrients abound. It surrounds itself with acidity and creates an environment suited to itself and other acidophiles. It provides a habitat for such plants and also competes with them. *Sphagnum* determines the kind of plants it competes with—shallow-rooted plants of low growth, too small to shut out the light. By holding water so effectively, it soaks the ground and deprives would-be competitors of oxygen.

The water-holding capacity of *Sphagnum* and *Sphagnum* peat affects soil temperature, nutrient availability, and gas balance. The uptake and retention of water vary with growth form and spongy structure. Water-holding capacities of living *Sphagnum* (175) range upward to about twenty-seven times the dry weight, but meaningful values for each species have yet to be determined. The reason is that species of compact growth are more spongy than those that grow loosely. Therefore, equivalent volumes of the plants as they function in nature need to be compared in weight, wet and dry. *Sphagnum* acidifies the environment by taking mineral cations from solution and giving back hydrogen ions. In this way it initiates the changes that transform fens into open bogs and eventually spruce muskegs. The species of *Sphagnum* have their own habitat preferences and capacities for cation exchange, those closest to the groundwater source having the least and those farthest from it having the most (10, 240, 260). Some grow in mineral-rich zones, others in older parts of the mat where minerals are hard come by. Cation exchange gives *Sphagnum* the means of extracting minerals from dilute solutions. The species arrange themselves in gradients according to their ion-exchange and water-holding capacities and are, for that reason, good indicators of abiotic changes.

Near the leading edge of a mineral-rich sedge mat, *Sphagnum teres* and less commonly *S. subsecundum* initiate the acidifying process. The wet *Sphagnum* lawn species are *S. cuspidatum* and sometimes *S. majus* in soaks and puddles, with *S. papillosum* in carpets just above water level and *S. capillifolium* and *S. magellanicum* initiating hummocks. The hummock formers are eventually, in the bog stage of development, topped by *S. fuscum*. When the hummocks reach maturity, the hollows are relatively dry. *Sphagnum recurvum* may fill such hollows as a loose carpet. It is encroached on by the hummock formers and often continues upward growth in mixture with them. The dominant species of spruce muskegs are *S. magellanicum* and *S. recurvum*. In the dense shade of a wet lagg *S. teres* and *S. subsecundum* may grow with cedar swamp species of *Sphagnum*. Cedar swamps (which sometimes encroach on laggs) have a calcium-tolerant *Sphagnum* flora consisting of *S. centrale, S. wulfianum, S. warnstorfii, S. girgensohnii,* and *S. squarrosum. Sphagnum fimbriatum* and *S. russowii* are common in cedar swamps but not restricted to them.

In calcareous fens, *Sphagnum warnstorfii* grows in brilliant red-purple mounds, and *S. fuscum* forms compact mounds, higher and rich brown. *Sphagnum warnstorfii* in cedar swamps has a less compacted growth form and a dark-green color. *Sphagnum fuscum* occupies the most acid niches in bogs. Its success in rich fens depends on its outstanding ability to acidify its surroundings. The top of a mound can have a pH of about 4 in contrast to 7 or more in the wet surroundings. This amounts to a thousandfold difference.

Sphagnum hummocks commonly grow up around tussocks of sedges or lower branches of shrubs. *Polytrichum strictum* may also provide such support. This moss grows apace with the hummocks and becomes a conspicuous feature at their tops. Its dense clumps of long, woolly stems may aid in the upward movement of capillary water.

Theoretically, when a hummock reaches its maximum size, it begins to deteriorate, becoming a hollow surrounded by new hummocks developed by successional stages from hollows (189, 269). Thus, the bog surface is continually involved in a regeneration cycle. However, the actual existence of such cycles may be questioned (14, 265). The water table continues to rise as peat accumulates, and in oceanic areas the hummocks may not dry out to the extent of initiating deterioration. They become hollows not by death and decay but by upgrowth of surrounding hummocks. Hummocks and hollows merely replace one another. Thus, in domed bogs of rapid peat deposit, continual regeneration may result in arcs of *Sphagnum fuscum* peat in alternation with paler peat. But such arcs may not be demonstrated in blanket bogs because of greater decomposition associated with the ionic burden of ocean winds and aeration provided by slope drainage. Bog surfaces in Great Britain show little evidence of hummock-hollow regeneration cycles but show instead a number of dry periods over the past thousand years when peat became humified followed by wet periods when peat again accumulated above "recurrence surfaces" (14). On the basis of historical records and stratigraphic evidence, the wet intervals have been dated at A.D. 900–1100, 1320–1485, and 1745–1800. In areas of continental climates and greater water stress, the water table is periodically halted in its rise by dry seasons or drought cycles, and hummocks are more likely to deteriorate to a hollow level and then regenerate. But the peat accumulates too slowly there to give evidence of hummocks becoming hollows and hollows becoming hummocks in a continuous regeneration cycle.

Bog Acidity

Bog waters are low in electrolytes and acidic in reaction. The acidity can be attributed to a number of factors (33). Because the water is not or poorly buffered, the pH is very sensitive to the dissociation of acids, even weak ones like carbonic acid. The pH is generally somewhat lower than 4, but carbonic acid dissociates down no further than 4.4. Sulfite, sulfate, chloride, nitrous, and nitric ions brought in by precipitation reduce the pH to about 3.6. (Sulfate ions are especially important. The predominant ions of bogs are, in fact, H^+ and SO_4^{--}, the makings of sulfuric acid.) Other sources of acidity lower the pH to 3.3 or somewhat less: Some acidity can be attributed to the cation-exchange activities of *Sphagnum* and *Sphagnum* peat (32, 34), and the dissociation of weak organic acids resulting from anaerobic decomposition may be an even greater source of acidity (fig. 5-1) (128). (Humic substances—humic and fulvic acids and humin—make up the bulk of the organic matter in older layers of peat. These substances result from the tardy decay or chemical transformation of proteins, fats, cellulose, and lignin followed by the synthesis of yellow-brown substances of high molecular weight.) Galacturonic acid bound to cell walls of *Sphagnum* contributes acidity by dissociation, as well as by ion exchange.

Cation exchange involves the adsorption of positively charged ions to the surface of negatively charged soil or humus particles. The adsorptive properties of humic substances far exceed those of clay. Peat added to a mineral soil contributes immeasurably to the ability of the soil to accumulate and retain nutrient cations. Humic substances are also effective chelators, holding cations by chemical binding rather than adsorption. Complexing by chelation holds metal ions that might otherwise be in short supply and releases them gradually. Peat has a very high cation-exchange capacity. Peat formed under mineral-rich conditions, as in fens, has a high degree of base saturation, but peat formed under cation-deficient conditions, in bogs, is saturated with hydrogen ions available for exchange.

The high ratio of organic matter to minerals in fen and bog peats has profound chemical consequences. In mineral soils most of the exchange sites are saturated by metallic cations, but as organic content increases, exchangeable hydrogen ions increase. As fens give way to bogs, exchangeable mineral cations diminish, and hydrogen ions dominate. In ombrotrophic situations, where mineral ions are in scant supply, about 10 percent of the cation exchange capacity may be taken up by bivalents, a small percentage by monovalents (other than hydrogen), and the rest by hydrogen ions.

Ions attracted to soil surfaces are held in equilibrium with those in the soil solution and can be replaced, or exchanged, in response to changes in concentration. As water percolates through the soil, some ions leave soil surfaces to balance those in the soil water. Cation exchange can, in this way, be reversed. For example, as Ca^{++} is removed by the roots of plants or flushed away by water, its place can be taken by H^+ ions until the soil is too acid for plant growth. A soil solution containing calcium and sulfate ions becomes a solution of sulfuric acid when hydrogen ions are exchanged for calcium. The more H^+ ions that are added, the more cations of other kinds are displaced and leached away. In a bog, the older peat is impermeable to water and therefore unresponsive to rain flush.

Hydrogen ions, held by one valence, are readily displaced by those of higher valence, and bivalent ions such as calcium, magnesium, zinc, and copper are also displaced by ions of higher valence. Iron can be retained with a force of two or three valences, manganese with two or four (or more). Obviously the greater the valence the tighter the bond and the more difficult it is for plants to make use of adsorbed ions. However, flushing by rain and snowmelt makes such ions available, and acid rain causes cations to be removed from exchange sites owing to reaction with pollutant anions. The cations accumulated by exchange sometimes exceed those in solution by 100 times. Such an accumulation makes survival possible in habitats otherwise nutritionally marginal. It is because of preferential adsorption that the bivalent ions of calcium and magnesium are more abundant in bogs than the acidity of the environment would suggest. That is also the reason that monovalent potassium and ammonium quickly leach away.

Sphagnum has a high capacity for retaining cations because its empty water-storing cells present a large surface for adsorption and also because of a high concentration of unesterified polyuronic acids (40, 240) bound to the cell walls (making up as much as 20–30 percent of the dry weight). Such long-chain carbon compounds as galacturonic acid are similar to sugars in structure except that every sixth carbon is part of a carboxyl group, -COOH. The hydrogen of each carboxyl group can be exchanged for a cation. Galacturonic acid and exchangeable hydrogen are continually produced at the growing tips of *Sphagnum* plants. In this way, living *Sphagnum* serves as a continual source of acidity. The high surface area and ion-exchange capacity of *Sphagnum* are long retained in fibric peat, but no new galacturonic acid is produced, and so peat is less effective in exchange, over time, than actively growing *Sphagnum*.

The capacity for cation exchange is not unique to *Sphagnum* (or *Sphagnum* peat). It is high as compared with higher plants, yet other mosses have capacities one-half as great, or more. However, other mosses do not grow in the density of *Sphagnum*, and they have only minor significance in peatlands. The ability to maintain acid surroundings while rain flushes the *Sphagnum* mat with a solution of minerals, however dilute, depends on the continued production of exchange sites. Hence, there should be greater acidity at the season of maximum growth, and that is the time when mineral flushing is also considerable.

Moore and Bellamy (152) compared summer and winter pH values in a *Sphagnum* lawn community in Britain: *Sphagnum cuspidatum* at groundwater level varied from a September value of 5.0 to a January reading of 5.1, in contrast to *S. rubellum*, a mere 2.5 inches higher, with summer values of 3.1–3.4 and winter values of 3.5–3.7. These values demonstrate a greater release of H$^+$ ions in the growing season. Another comparison of winter and summer values, at two British localities, reported by Rose (211), shows considerable differences, according to the season and vertical position, from hummock to hollow (from *S. rubellum* downward):

	Sept. 22	March 24		Sept. 9	Jan. 7
S. rubellum	3.8–3.9	4.6	*S. rubellum*	4.5	5.7
S. magellanicum	3.9–4.6	4.9	*S. tenellum*	4.4	5.4
			S. papillosum	4.5	5.7–5.9
S. tenellum	4.4	4.8	*S. palustre*	4.1	5.2
S. papillosum	4.0–4.2	4.7	*S. pulchrum*	4.9	5.9
S. pulchrum	4.9–5.0	4.8	*S. recurvum*	4.1	5.8
			S. cuspidatum	5.8	5.2

The differential capacity of *Sphagnum* species to adsorb cations is indicated by pH differences from the bottom to the top of hummocks (fig. 5-5) (10, 152, 260). This differential is illustrated also by higher concentrations of pollutant ions, lead, zinc, and manganese, held by *Sphagnum fuscum* at hummock tops as compared with species of the *Cuspidata* in hollows (179). In highly ombrotrophic peatlands of Finland (176), *Sphagnum fuscum* holds considerably more calcium ions than the *Sphagna* of hollows do, apparently because of a greater exchange capacity at hummock tops. Other compact-growing hummock species, such as *S. capillifolium*, also have a heightened capacity for trapping cations. But it is more than density that makes hummock species capable of greater ion exchange. The higher the species grows above a mineral water source the higher its concentration in galacturonic acid and the higher its exchange capacity. Cations accumulated in hollows by downwash also contribute to the bottom-to-top gradient. There is a lesser capacity and a lesser need for ion exchange in the hollows, where minerals are relatively abundant.

It has been recorded, in Britain, that *Sphagnum subnitens* invading a calcareous fen with soil water at pH 8.3 had a pH of 6.5. Even greater differences can be detected in Michigan sedge mats acidified by *S. teres* in some cases no more than a yard away from mineral-rich water. Hummock-hollow contrasts are also striking (fig. 5-5). In northern Michigan, readings at the top of hummocks show a greater acidity than those in the hollows, at the level of groundwater, ranging from 2.7 down to 4 (260). (It is likely that these readings are too low, but the degrees of difference are significant.)

Reed-sedge deposits, resulting from fen vegetation, are generally well decomposed. Bog deposits show, below a surface layer of living *Sphagnum*, an accumulation of loose, fibric peat overlying more decayed and compacted hemic and sapric peat, with the bottom layers commonly consisting of sedge or sometimes forest peat. Sometimes the deposits show sedge or *Sphagnum* peat in alternation with lake sediments, such as marl or gyttja, owing to a history of submergence. The surface peat, below the living *Sphagnum*, may have a pH of about 3.1–4.5. The lower peat gives less acid readings owing to the kind of vegetation and the water chemistry at the time of deposit, whether conducive to bog or fen growth and favorable to decomposers or not. The most abrupt change in pH readings occurs at the stratum at which *Sphagnum* entered the vegetational succession and became a component of the peat. Readings from a British deposit

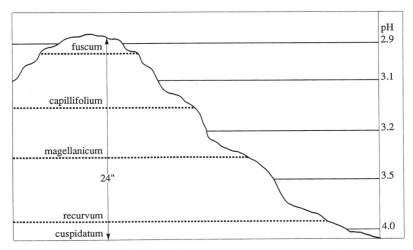

Fig. 5-5. The pH gradient as demonstrated by species of *Sphagnum* in hummock-hollow complexes in northern Michigan. (Redrawn from Vitt, Crum, and Snider 1975.)

have shown a surface pH of 3.61 in contrast to 4.68 at a depth of 13 feet. At Dingman Marsh in Cheboygan County, Michigan, readings between 14 and 58 inches showed a pH spread from 3.1 to 5.1 and at a different site, between 3 and 103 inches, a spread from 4 to 5.2 (93).

Acid Rain

Rain is normally somewhat acid, about pH 5.6, but in the last three decades it has become much more acid, at least in areas subject to atmospheric pollution. In Scandinavia, pollutants brought in by air currents from Great Britain and the Ruhr Valley have increased the acidity of precipitation by 200 times during the past decade. In large areas of northern Europe the pH of rain is less than 4, and values less than 3 have been measured: a storm in Scotland in 1974 brought in rain of pH 2.4—an acidity greater than that of vinegar and of bog waters at their extremes.

An increase in acidity in the northeastern United States can be traced, in large part, to sulfur dioxide pollution produced by the coal- and oil-burning industries of the Ohio Valley. Sulfur dioxide goes into solution in atmospheric humidity as sulfurous acid and, in greater concentrations, as sulfuric acid. Acid rain and snow cause nutrients to be leached away and also increase the solubility of toxic minerals such as zinc, copper, aluminum, and manganese. In areas of acid rock and soil in New England, forests have been killed and lakes depleted of fish. Lakes with pH levels of 4.5–6.5 are highly sensitive to acid input by rain and snow, and small lakes in mountain watersheds are particularly vulnerable. An abrupt increase in acidity takes place every spring as snow melts. The meltwater sometimes has a pH of 3, and the lakes may have their pH levels lowered to about 4 at the time that many fish spawn. Nutrient cycling is reduced because of lessened activity of decay bacteria. This causes plankton to die of mineral deficiencies and fish dependent on plankton to starve. The control of pH depends on carbonate-bicarbonate buffering. As the water is depleted of bicarbonate and sulfate anions become dominant, such pollutants as aluminum and lead seriously affect the biota. Where soils are calcium containing, in the Midwest, for example (127), lakes may become more alkaline because acid precipitation increases the rate at which soils weather and calcium bicarbonate drains into lakes. But in areas of acid rock, lake waters are poorly buffered and become more acid.

Acid rain does not favor the growth of *Sphagnum*. Even in low concentrations sulfur-containing anions, and particularly bisulfite ions, HSO_3^-, reduce the growth of *Sphagnum*. The addition of pollutant anions can cause chemical combinations that contribute to a loss of cations which are, at best, scarce in *Sphagnum*-dominated boglands.

Water Relations

Sphagnum is admirably suited to taking up and holding water, yet it suffers from occasional dry spells in summer and water stress occasioned by temperature extremes, cold and hot, diurnal and seasonal. The uptake and distribution of nutrient ions are close-linked to water movement. Although some *Sphagna* move water efficiently, the hummocks are broken up by roots and leaves and are also compacted and supersaturated within and therefore inefficient in capillarity. Therefore, a prolonged drop in the water table may kill the *Sphagna* of hummock tops. Both flooding and drying endanger the survival of *Sphagnum*. That is why peatlands develop in basins of relatively constant water supply rather than along watercourses with spring highs and summer lows. And that is why *Sphagnum* can invade floating sedge mats that accommodate to seasonal changes in water level.

The absorption, retention, and movement of water are related to the spongy construction characterizing the genus, but they vary in degree with the species. The variations are related to compactness of growth, cortication of stems by pendent branches, imbrication of leaves, and cellular differentiation. Water is sucked up and held by large, empty cells of stems, branches, and leaves. The thin walls of water-storage cells of leaves are strengthened by ringlike thickenings. Those of leaves and in many cases those of stem and branch cortex are provided with surface pores. Slow-decaying, *Sphagnum* retains its spongy character long after becoming peat.

A loose growth form characterizes the submerged *Sphagna* of drainage soaks and the carpet species of depressions, in contrast to the density of hummock formers. Owing to their cellular structure, the species of hollows, like *S. cuspidatum* and *S. recurvum*, hold less water than those of hummocks, such as *S. magellanicum* and *S. capillifolium*. Although they dry out more quickly with the onset of summer heat and drought, the species of hollows survive drying well and recover quickly.

In order to obtain sufficient nutrition, *Sphagnum* needs to absorb and conduct as much water as possible. Warmed by the sun, the species at hummock tops have the greatest evaporation stress and also the greatest capacity for capillary movement and ion exchange. They are therefore able to make good use of ions in dilute solutions brought up by capillary flow. Although the hemispherical form of hummocks presents relatively little surface for water loss by evaporation, the hummocks are occupied by red and brown *Sphagna* that absorb heat and increase the solute movement by evapotranspiration pull at the same time that they contribute microclimates favorable to metabolic functions over longer seasons. *Sphagnum* of low, compact growth form copes well with cold and dry extremes, because it traps and retains heat and moisture. The hummock growth guards plants rooted in *Sphagnum* from diurnal fluctuations in temperature and unseasonable frost.

Chapter 6

Peatland Archives

> If you have built castles in the air, your work need not be lost; that is where they should be. Now put the foundations under them.
> —Thoreau

Once warmth prevailed throughout the Northern Hemisphere. In the far distant past, a fern flora deposited coal in the Arctic, and later a warm-temperate vegetation left fossils there similar to those deposited in the late Mesozoic and early Tertiary far southward around the Mississippi Embayment. The wide-ranging, uniform distributions of those plants were dissected by mountain formation ushering in the Tertiary and associated changes in wind and rain, and arctic climates were altered. As the Tertiary wore on, climates became cooler and dryer. Eventually, in the Pleistocene, snow no longer melted away during the short summers of the north. Rather, it accumulated year after year for long ages and compacted as a heavy burden of ice in subarctic and montane areas of North America and Europe. As the ice cap increased in depth and extent, pressures at the center caused pushing, gouging, scraping, and scouring at the margins. A series of glacial advances and retreats modified the topography in eastern North America as far south as the Ohio River and altered climates far south of that. Northern forests containing spruce, white cedar, tamarack, and pine—and the hairy mammoth—moved as far south as the Gulf of Mexico.

And then the climate moderated, and glacial ice began its final retreat. Torrents of meltwater dumped sand, gravel, and boulder clay in a pattern of disarray, leaving outwash plains and drainage-blocking moraines, kames, and eskers. Thousands of lakes, large and small, resulted from basin scouring in solid rock, damming by glacial landforms, and melting ice blocks buried in drift. Wave action in large, postglacial lakes formed sand bars and spits and created quiet-water embayments, some of which persisted long after the parent lakes disappeared. Onshore winds piled up dunes in alternation with wetland depressions. As the glacier melted, the land surface, free of its weight, rebounded, tipping water into new patterns of drainage. The present area of the Great Lakes was delimited, and lowlands with water tables at or near the surface became available for paludification and bog formation.

First a tundra vegetation and then forests of spruce, fir, and jack pine moved into the postglacial barrens and contributed pollen to the lake sediments. The pollen deposited year after year, century after century, provides a continuous, datable record of climate and vegetation. Some of the lakes, very many of them, were encroached on by vegetation in fen-to-bog sequences and gradually filled in with peat, as did other basins originally lacking water. In the north, around the shores of Lake Superior, the peat deposits are generally not deep, because the lakes are not old. But farther south, in northern Indiana, for example, lakes have had time to fill up with as much as 70 feet of peat and other sediments. In the Myrtle Lake region of northern Minnesota the peat is about 26 feet deep (103). In Maine peat deposits are, at most, 42 feet deep (281), in the Avalon Peninsula of Newfoundland 20 feet (186). In Quebec, in the St. Lawrence River valley, peat may be more than 30 feet deep; north of that, at 46° N about 24 feet; at 48° about 15 feet; at 51° about 9 feet; at 55°, the forest-tundra transition, about 6 feet; and in the tundra of northern Ungava seldom in excess of 2.3 feet (198). (In Great Britain, blanket bogs have a peat accumulation of 6 feet or more, and raised bogs as much as 30 feet [152].)

Plants and animals, now as in the past, live where they can, where the climate is suitable. If the climate changes, they move out or die out. The evidence of change can be read in lake mud and peat deposits. A bog's history can be reconstructed from the color and texture of the peat and the macrofossils—fruits, seeds, and leaves—buried in it. But the microfossils, especially pollen grains, serve as a less localized, regional record of vegetational shifts in response to climatic changes. (One gram of peat usually contains some 50,000 wind-deposited pollen grains.) Small fens and bogs carry a heavy burden of local tree pollen, but larger peatlands, such as the ombrotrophic mires of Britain, have a sparse cover of trees and provide a less distorted view of the regional past. Likewise, the muddy sediments of larger lakes give a better account of regional chronologies than small, peat-filled lakes do.

Limnologists, concerned with lake history, with chemical and biotic changes, work with lake muds rather than peat deposits. They reconstruct the history of the aquatic environment, from youth to senescence, making use of diatom frustules, sponge spicules, sporozoan and dinoflagellate shells, and arthropod exoskeletons as evidence of mineral status, oxygen content, and productivity. Pollen correlates such sedimentary remains with the regional climatic history. Nutrient loading may be determined from nitrogen, phosphorus, and carbon assays, and carotenoids specific to certain kinds of algae provide an index to eutrophication. Changes in trophic status have been detected in a small lake near Rome, the Lago di Monterosi, which accumulated nearly 8 feet of sediments in 25,000 years, after which, in about 2,000 years, it added 4.7 feet (114). The reason for this considerable increase in sedimentation is that a road built in 171 B.C. blocked drainage and increased the nutrient supply and productivity of the lake. As a further example, Linsley Pond, in Connecticut (58), was formed in a kettlehole 13,000–14,000 years ago. Pollen preserved in the mud reveals regional climatic changes favoring a spruce-pine-oak sequence. The pine phase correlated with an abundance of a relatively large cladoceran presently found only in waters free of fish, in other words, in oligotrophic waters. As the fish population responded to increasing eutrophy, that species was replaced by a smaller one less vulnerable to predation. In the same mud, a shift in midge larvae also showed a change from oligotrophy to eutrophy.

In small bog lakes it is easy to find the deepest part of a basin, where the fossil record is most complete. Samples can be taken in summer from the surface of a well-formed mat, but sampling through the ice may be easier than working from a tippy boat or on an unstable sedge mat hip-high in water. A Hiller sampler is among those commonly used for pollen analysis. It consists of an auger bit at the end of a cylinder enclosed in a sleeve that can be rotated to open and shut positions (fig. 6-1). An opening in the sleeve that can be made to coincide with an opening in the cylinder has, at one side, a bladelike edge such that a few clockwise turns will cut and store peat in the cylinder. Counterclockwise rotation closes the cylinder and keeps the peat free of contamination as it is extracted. A crossbar handle is used for rotating the closed cylinder. By use of extension rods the sampler can be used at depths up to 30 feet or more (fig. 6-2).

Pollen grains are resistant to almost anything but oxidation, and therefore peat samples must be kept wet (and anaerobic) until analyzed. A few ounces of peat taken from each sampling level can be loosened up by soaking in alcohol and boiling in potassium hydroxide or hydrochloric acid. A more refined procedure, *acetolysis*, now used routinely, involves heating in glacial acetic acid and concentrated sulfuric acid. The object is to get rid of cellulose, lignin, and humic substances and retain pollen grains. After sieving and centrifuging, the suspension can be stained with gentian violet or safranin to differentiate pollen from extraneous organic matter. A small amount of the suspension mounted in glycerin or silicone is examined at 200–500× magnification. (Oil immersion, giving 1000× magnification, may be used in critical studies.) Using a mechanical stage so that no part of the slide is observed more than once, 300 grains, or more, are identified. The percentages of each kind of pollen are charted according to depth. The resulting graph, or *pollen diagram*, gives indication of the wind-pollinated genera and species represented in the forests of the past and their changes in abundance. Marks of identification include shape, size, surface sculpturing, and number and location of pores. Identification is expedited by the use of reference slides, descriptions, and illustrations. Some pollens are difficult, if not impossible, to distinguish. Sometimes, as in the Umbelliferae, only the family can be recognized. Pollen grains of spruce, pine, and fir have saclike or winglike appendages with some difference in shape and attachment, but the size of jack pine pollen overlaps that of red and white pine, and so does the size of white and black spruce pollen. There is no way to distinguish species of oak or birch by their pollens.

Pollen studies in southern Sweden (92) show that a treeless tundra was replaced some 10,000 years ago by birch, a cold-climate indicator in European peats, and then by cold-tolerant pine. About 9,500 years ago the climate became warmer, and birch began to decrease in abundance while pine increased and hazel became abundant. By 7,700 years ago, a mixed deciduous forest of oak, elm, alder, hazel, and lime (*Tilia*) replaced pine and dominated under conditions of warmth for five millennia. Conditions were relatively dry until 2,500 years ago when the climate became abruptly cooler and wetter and soils became flooded or, at least, waterlogged. Elm, oak, and lime decreased in abundance, and spruce, beech, and hornbeam took on importance. This abrupt climatic change is marked by a thin, dark *boundary layer,* or *Grenzhorizont,* corresponding to a *recurrence surface* indicative of elevated water tables (associated with a rise in ocean levels caused by a tardy disappearance of glacial ice). It also corresponds to a climatic shift favoring peatland development.

Climatic Period	Nature of Climate	Forest Type	Time
Preboreal	Cool summers, severe winters	Birch	8000 B.C.
Boreal	Dry, warm, continental	Hazel, pine	7500 B.C.
Atlantic	Moist, warm, oceanic	Mixed oak	5600 B.C.
Subboreal	Dry, warm	Mixed oak; marked at the end by the Grenzhorizont	3000 B.C.
Sub-Atlantic	Cool, wet	Spreading beech, hornbeam, spruce	500 B.C.

In the British Isles, climatic and hydrological changes at the time of the *Grenzhorizont,* about 500 B.C., marked the end of the Bronze Age and the beginning of the Iron Age (92). The Roman conquest of Britain began in the Iron Age. A decrease in some tree pollens and an increase in herb pollen document Iron Age activities in forest clearance and agriculture. Continued wetness is demonstrated by the depth of peat overlying the *Grenzhorizont* and burying Roman artifacts, including coins dated as late as A.D. 410. Greek and Roman coins buried in Scandinavian peat among Viking relics show a similar chronology. With some fluctuations, relatively cool and moist climates have continued from then onward. A generalized pollen diagram from the fens and bogs of southeastern England (fig. 6-3) shows a vegetational and climatic sequence very similar to the Swedish (92).

In North America the timing of events is similar (57, 270), although allowance needs to be made for regional variations in climates as well as differences in the migratory history of species.

Some differences are to be expected, of course, in inland regions where wetness associated with changes in ocean levels were not a factor. Sears (226), before the time of radioactive carbon dating, worked out a generalized sequence of vegetation and climate in eastern North America:

I. Cool, moist; spruce-fir maximum; ca. 11,000 years ago
II. Warm, dry; pine maximum, with oak generally present; ca. 9,000 years ago
III. Warmer, more humid; beech maximum, with hemlock sometimes present; ca. 6,000 years ago
IV. Warm, dry; oak and hickory maximum; ca. 3,500 years ago
V. Cooler, moister; increase in maple, hemlock, chestnut, and pine; present time

Also, he reconstructed the climate of northern Ohio (225) on the basis of pollen studies:

Peat Depth in Feet	Climate
12–14	Cold, wet (as in northern Labrador)
7–11	Gradual shift from oceanic to continental climate
4–6	Cool, dry (as in southern Manitoba)
2–3	Maximum desiccation
2	Abruptly cooler and moister (as in the northern Great Lakes region)
1	Moderation of temperature with increased humidity (as in north-central Ohio at the present time)

The flora of the eastern United States has been minimized by vicissitudes, geologic and climatic, and more recently by the depredations of man. It is a fortuitous assemblage of species, hardy survivors of sixteen to eighteen glacial-interglacial cycles. On the final retreat of Pleistocene ice, beginning 15,000 to 16,000 years ago, a tundra flora occupied the vast area of obliteration. The pioneering species belonged to a fugitive assemblage occupying a fringe of wetness adjacent to the glacial front both in advance and retreat or crowded into a well-established and competitive flora far beyond the glacial front. Very likely the pioneers once belonged to a rich and diverse flora of the north. The crowding of plant and animal life that once occupied millions of square miles caused wholesale extermination owing to unsuitability of habitats and stiff competition. Among the surviving few, genetic diversity was diminished because of low population sizes and reduced gene pools.

Changing wind patterns and altered climates (including cool microclimates caused by meltwater drainage into the groundwater) made a southward advance possible at the same time that they put stress on the flora and fauna indigenous to the south. Both fugitive and resident species suffered. Camels, horses, mastodons, mammoths, giant sloths, and giant beaver vanished from the continent. Forest trees now northern in distribution ranged far southward. Pollen of *Thuja, Larix, Abies,* and *Picea* was deposited in sediments far south of the glacial front. In Texas and Florida, fossil pollens of *Abies* and *Picea* and assemblages of diatoms presently northern in distribution give evidence of environmental stress far southward. Buried cones of tamarack and white spruce and twigs of white cedar demonstrate that northern tree species indeed grew as far south as deposits of their windblown pollens suggest.

As Pleistocene ice moved to the north, the spruce and pine forests of the south were replaced by deciduous woodlands. Spruce replaced the postglacial tundra in the northeastern United States by about 12,500 years ago. Fir and jack pine occurred with it but became more abundant as climates moderated. About 10,000 years ago pine and fir assumed dominance. As spruce, fir, jack pine, and paper birch moved northward, so did moose, elk, caribou, and musk-oxen. Meanwhile, oak and other deciduous trees gained in ascendancy as a response to greater warmth. The climate became progressively warmer and dryer, but about 5,000 years ago species indicative of xerothermic conditions declined in abundance and contracted in range. About 2,000 years ago, under the influence of cooler and moister conditions, boreal elements began to increase in abundance, and this trend has continued to the present time.

Forest successions were slow to develop in the Great Lakes states because of unsuitable soil conditions. Soil profiles were scarcely developed in late glacial time because of solifluction and rapid erosion. The early postglacial vegetation consisted of both wet and dry tundra plants, as indicated by pollen and other fossil remains of *Salix herbacea, Dryas integrifolia, Vaccinium uliginosum* subsp. *alpinum,* and *Selaginella selaginoides.* Sedge pollen was deposited in abundance. Alder invaded later but progressively increased in abundance after its arrival. The wetland invaders were presumably those that lived near the glacial front and could thrive in similar newly available fen habitats, cold, wet, and mineral rich. The assemblages of species presently found in rich fens along the shores of the Great Lakes are similar to those of the postglacial fossil floras as well as modern assemblages far northward in arctic and boreal latitudes. Fens and marshes no doubt preceded bogs and swamps, as they do in modern vegetational successions. The upland invaders of postglacial times were plants that could occupy soils of little or no organic content. Upland forests developed slowly in patterns of succession dependent on climate, soil maturation, and biotic interaction. The associations of species, both plant and animal, came slowly owing to differences in migration rates. Habitat extremes in temperature, drainage, and mineral nutrition set limits on the kind of plants that could pioneer, and diversity diminished at each stage of succession. Little wonder that the peatlands of glaciated regions had and still have an essential uniformity over a broad northern range and an extremely limited diversity of species in their more mature manifestations.

At the time of European settlement, spruce and fir prevailed in the boreal forest, and deciduous forests of sugar maple, beech, and hemlock characterized the better soils of the upper Great Lakes region, with red and white pine common on sandy soils. In the lower Great Lakes region, oak, hickory, and chestnut were dominant. The forests have been drastically altered since that time by lumbering, agriculture, and industry. The pollen record shows a dramatic decline in forest trees and an increase in grasses and weeds of arable soils.

Pollen studies in Michigan peat show a spruce-pine-hardwood sequence. The pollen deposited at Hartford Bog (286), Van Buren County, southwestern Michigan (fig. 6-4), shows that a cool, moist spruce-fir period was followed about 8,000 years ago by a warmer and drier climate under which pine flourished in association with oak, hickory, and other broad-leaved trees. But about 2,500 years ago, the climate became cooler and moister, and the forest became a depauperate oak-pine association. It was this forest that fell to the axe as the frontier moved westward during the past century. Also in southwestern Michigan, at Froelich Bog in Berrien County (fig. 6-5), a dominance of gymnosperms is revealed at peat depths of 4.5–13.5 feet and a later dominance of broad-leaved trees at 0–4.5 feet. Carbon-14 dates of 11,940 years ago were established for the height of the gymnosperm period and a range of 5,450 to 3,710 years before the present for the middle of the deciduous tree period.

Within Michigan the time of glacial retreat varies from 14,800 years in the south to 11,800 years in the north. A chronology for southern Michigan shows the time spread from spruce domination 13,200 years ago to the present oak-pine forest dominance established 2,500 years ago. Of special interest is the warm, dry period from about 4,080 to 2,500 years ago when the prairie expanded eastward into the lower Great Lakes region, as a "prairie peninsula" and as scattered grasslands, or "oak openings." (These grasslands were quickly taken up by settlers because of easy clearing and soils improved in humic content by grass cover. James Fennimore Cooper lived briefly in Schoolcraft, in southwestern Michigan, where such disjunctive prairies are found, and, while there, wrote one of his less known novels, called *Oak Openings*.)

Stratigraphic investigations in northern Michigan reveal a progression from spruce-pine to pine dominance with an increasing abundance of such northern hardwood elements as maple, beech, birch, and hemlock dating from 5,000 years before the present. At Lake Sixteen, in Cheboygan County (74), lake-fill began as early as 6,200 years ago. Climatic changes about that same time caused an elevation of water tables and eventually resulted in paludification external to the lake basin nearly 3,000 years after infilling began. Also in Cheboygan County, pollen stratigraphy (148) reveals that Gleason Bog was a shallow pond supporting a rich fen vegetation about 8,000 years ago. After being partially filled with peat and sand, about 4,000 years ago, it changed to an oligotrophic bog and is now completely filled in by a *Chamaedaphne-Sphagnum* vegetation. At Gates Bog, nearby, paludification starting 3,800 years ago caused peat to accumulate over a sandy depression without an initial pond phase. The onset of peat accumulation at both sites is attributed to a rising water table associated with the onset of cool and moist climates.

Along the Minnesota-Ontario border, pollen deposits record cool or cold, moist climates that became warmer and drier before changing to the relatively cool, moist climates of the present. Spruce-fir or spruce–fir–jack pine was the initial forest cover. It was abruptly replaced by jack pine eventually grading toward warmth as a jack-red-white pine cover and finally into a pine-spruce-fir mixture indicative of cooler conditions. Warm-climate oak, elm, hemlock, walnut, and hickory had some representation, particularly during the lengthy dominance of jack pine. Birch had some presence throughout, probably owing to fire rather than cold. Potzger analyzed fifteen bogs in the region (194, 196). One of them, Johnson Camp Bog (fig. 6-6), showed the shift from spruce-fir to jack pine and the beginning of a warmer trend at the 8-foot level, carbon-14 dated about 7,100 years ago.

Since the early pollen work detailed above, Michigan chronologies have been firmed up by additional paleoecological studies and carbon-14 dates. In the absence of carbon-14 confirmation, it is possible to assign dates to certain marker species as their migration patterns are known and their time of entry into the pollen record can be inferred. Just as the hemlock decline of 4,800 years ago is a good marker, the recent decline of chestnut (owing to chestnut blight) and elm (owing to Dutch elm disease) will serve as markers in future time. The settlement period of the past century, with its sharp increase in weed pollen associated with agriculture, also provides a means of marking time.

Archeological Remains

European peats have yielded many artifacts of the period from about 800 B.C. to A.D. 400 (92). This was a time of transition from the Bronze to the Iron Age and the time of agricultural settlement. Roman invasion and colonization are nicely represented in such deposits. As many as 2,000 "bog bodies" have been found in Europe in addition to the famed Tollund man of Denmark and the Lindow man of England. Many of the bodies give evidence of sacrificial murder or execution. It may be that the bog people belonged to a relatively sedentary type of society that had reason to exploit bog resources, but in North America, at least in the glaciated North, the Indians in their hunting and fishing pursuits, as in their slash, burn, and move agriculture, had little need to frequent peatlands where they could have been buried, accidentally or otherwise. In central Florida, an Indian burial site covered over with 10 feet of peat has yielded up bodies of more than 100 men, women, and children dating back 7,000 to 8,000 years ago (132). Woven and twined vegetable fabrics of surprising fineness were found, together with fabric-making instruments such as hooks, awls, and bone needles. The bodies showed a degree of cellular structure still intact and, in fact, brains with identifiable DNA. (The preservation, in marked contrast to the European bog people, which showed no signs of DNA, can perhaps be accounted for by burial in an oxygen-free though less acid peat.) The presence of old people as well as young, handicapped persons among the bodies suggests a society organized well enough to exist above a mere survival level and to support the more needy of its members.

It is possible, of course, that when North American peats have been dug into as much as the European the remains of man and his day-by-day effects will prove to be more numerous.

Fig. 6-1. A Hiller peat borer, also known as a Swedish borer, with its peat-collecting chamber open

Fig. 6-2. Barbara Madsen, paleoecologist, using a Hiller peat borer at Nolten Fen, Cheboygan County, Michigan

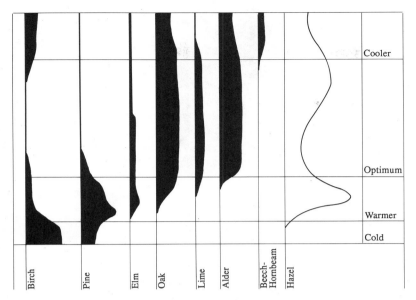

Fig. 6-3. Climatic and vegetational changes as shown by pollen accumulated in peat deposits in southeastern England. The chronology is similar to that demonstrated in Scandinavian peatland deposits. (Adapted from Godwin 1981.)

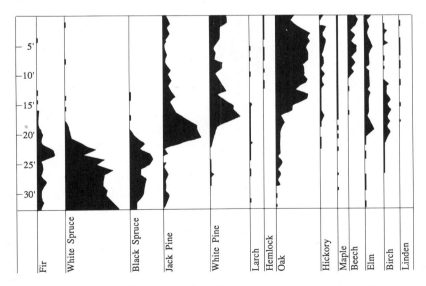

Fig. 6-4. Vegetational shifts in postglacial time as shown by pollen analysis of Hartford Bog, near South Haven, southwestern Michigan. (Adapted from Zumberge and Potzger 1956.)

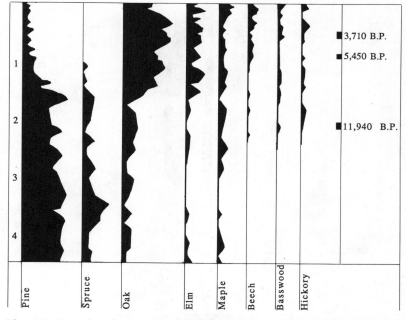

Fig. 6-5. Postglacial vegetational shifts as shown by pollen analysis of Froelich Bog, Berrien County, southwestern Michigan. (Adapted from an unpublished profile prepared by P. Ahern.)

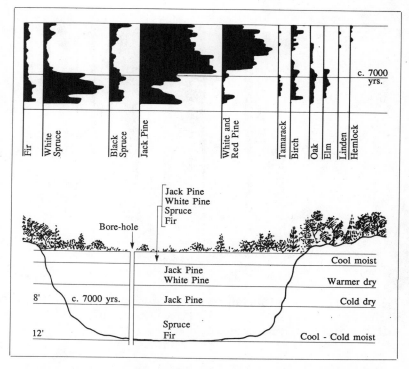

Fig. 6-6. A pollen diagram and bog section showing climatic and vegetational changes in northern Minnesota. (Adapted from Potzger 1950.)

Chapter 7

The Nature and Use of Peat and Peatland

> ... making the earth say beans instead of grass,—this was my daily work.
> —Thoreau

Peat, according to one of several definitions, is a soil formed under water-soaked, anaerobic conditions, accumulated to a depth greater than 30 centimeters drained and 45 centimeters undrained and having an organic content of at least 20 percent and an ash content of less than 50 percent. Its physical and chemical properties are based on its content of organic matter, minerals, gas, and water, and that content varies with vegetational origin and degree of decomposition. Peat can be classified according to vegetational origin (aquatic, sedge, wood, herbaceous, or *Sphagnum*) or degree of decomposition (fibric, hemic, or sapric). The principal components of peat, in addition to water, are cellulose, lignin, and colloidal humic substances resulting from their incomplete decay. Lignin is defined only as the substance left after treatment with strong sulfuric acid. *Sphagnum* peat in various degrees of decay may contain 20–30 percent of lignin, whereas sedge and forest peat may have 50 percent. In general, the lignin content increases with decomposition because of its resistance to microbial action. Humic substances consist of humic and fulvic acids in addition to an ill-defined humin, which is the fraction of organic matter not dissolved when peat is treated with dilute alkali. *Sphagnum* peat contains only 5–20 percent of humic acids in contrast to more decomposed sedge peats with 20–50 percent.

A useful classification of peat is based on degree of decomposition. Fibric peat, consisting primarily of *Sphagnum* moss, consists of more than two-thirds recognizable plant "fiber." This is the tan to light reddish brown peat of horticultural value. Hemic peat, or reed-sedge peat, is dark reddish brown and contains one- to two-thirds plant fiber. It is derived from sedges, reeds, rushes, and woody plants. Sapric peat is very dark and contains less than one-third plant fiber. It is derived from lake bottom sediments and from greatly decayed peat.

Peatlands are often drained for agriculture and exploited for horticultural peat moss (essentially undecomposed), potting soil and lawn dressing (greatly decomposed), and fuel (highly degraded and compacted). The extent of degradation, or humification, can be measured by fiber content, bulk density, color, water retention, and ash content.

The *von Post humification scale* (186) gauges the degree of decay from the top of the peat deposit downward, through successive layers of decomposition and vegetational change.

Apparent Humification (H)	Nature of Water Passing	Amount of Peat Passing Through Fingers on Squeezing	Residue after Squeezing
H-1 none	clear, colorless	none	unaltered
H-2 slight	clear but yellow-brown	none	almost unaltered
H-3 very little	turbid, brownish	none	slightly altered
H-4 little	very turbid, brown	none	remains somewhat mushy, difficult to identify
H-5 ±evident	muddy, rather dark	very little	very mushy, remains ±identifiable
H-6 ±evident	muddy, rather dark	one-third	very mushy, remains not distinct, scarcely identifiable
H-7 strong	very muddy, dark	one-half	very soupy, remains scarcely identifiable
H-8 strong	thick and gruelly	two-thirds	very soupy, remains scarcely identifiable, consisting mainly of more resistant root fibers, etc.

Apparent Humification (H)	Nature of Water Passing	Amount of Peat Passing Through Fingers on Squeezing	Residue after Squeezing
H-9 almost completely decomposed	thick, gruelly, no water	almost all, as uniform paste	little or no plant remains, homogeneous
H-10 completely decomposed	thick, gruelly, no water	all	amorphous, no plant remains

Horticultural peat is made up of light brown, slightly humified (H-1 to H-2) *Sphagnum* (more than 80 percent by volume) and an underlying, darker layer of some humification (H-3 to H-5) representing a minerotrophic-ombrotrophic transition and consisting of sedges in mixture with *Sphagnum*. Below that (H-5 to H-10) is a dark brown, stringy deposit suitable for fuel and consisting of more than 50 percent by volume of sedges. Fibric peat has humification values of H-1 to H-3, hemic peat H-3 to about H-5, and sapric peat about H-5 to H-10.

The von Post field method is easier to apply when reduced to five stages of humification. The von Keppeler method (186) based on weight percentages can be expressed as:

$$r = \frac{(I - II)100}{E(I - w + a)}$$

where r = % decomposition
I = weight of dry residue
II = weight of residue on burning
E = weight of air-dried sample
w = moisture content of air-dried sample
a = ash content of air-dried sample

The von Post scale can be correlated with von Keppeler's r values (186) as follows:

H-1–3	slightly humified, light brown	20–37%
H-4–5	slightly humified, medium-colored	38–48
H-5–6	rather strongly humified, brown	49–55
H-6–7	strongly humified, dark brown	56–64
H-8–10	very strongly humified, dark brown	65 and more

Fiber content correlates well with water content, retention, and movement. The fibrous nature is measured as the proportionate amount of plant fragments larger than 0.15 millimeters. In order to determine fiber content, a known weight of oven-dried peat is sieved under a gentle flow of water. The fibrous material remaining on the sieve is then dried, weighed, and expressed as a percentage of the original weight. If the peat is highly decomposed, fibers are nearly absent. But if it is only moderately decomposed, the fibers may be largely intact but easily broken down by handling, or rubbing. Therefore, the percentage of material that does not break down on rubbing gives a realistic measure of decomposition. The rubbed fiber content is determined by firmly rubbing that material eight to ten times between thumb and forefinger before computing the proportionate weight. Fibric peats have an *unrubbed fiber content* of more than two-thirds, hemic peats one- to two-thirds, and sapric peats less than one-third. The *rubbed fiber content* of fibric peat is greater than 40 percent, hemic peat between 17 and 40 percent, and sapric peat less than 17 percent.

Ash is the noncombustible mineral content. Peat formed in a mineral-rich fen has a higher ash content than that accumulated in a bog. The greater the decomposition, the greater the ash content. Thus, the amount of ash in lower levels of peat is greater than that in upper levels. The mineral constituents of peat are in part incorporated in plant remains and in part derived from extraneous sources such as dust or influx of mineral-holding water. The values can range from about 1 percent of the dry weight in fibric material to more than 50 percent in highly decomposed peat under cultivation.

Peat soils are lightweight and readily distinguished from mineral soils by *bulk density*. Bulk density measured in grams per cubic centimeter increases with decomposition and with the accumulated weight of overlying organic deposits. Thus, fibric peat has a density less than 0.09; hemic peat 0.09–0.2; and sapric peat more than 0.2. Bog peats, consisting mainly of partially decomposed *Sphagnum*, have a low bulk density and a high moisture content. Fen peats, being more decomposed, have a higher density and lower moisture content. The peats of rich fens are significantly higher in bulk density and lower in water content than those of poor fens.

Peat accumulated in sedge meadows in the southern part of the Great Lakes region may hold about 2.7 times its dry weight in water, but only 1.7 times when degraded to muck. Northward, where fen peat may be less decomposed, the water-holding capacity may be 3.2 times the dry weight as contrasted to 5.8 times in the peat of *Sphagnum* bogs. Pore size determines the amount of water retained by peat soils. Undecomposed peat contains many large, easily drained pores, whereas more humified peat with fine, less connected pores retains water. Hence, lowering the water table for agriculture drains a great deal of water from fibric upper layers of peat but very little from lower hemic and sapric layers. Compressibility, also related to porosity, is very low in sapric peat but high in fibric peat. (A horticultural grade of peat moss, consisting of scarcely decomposed *Sphagnum* with woody inclusions screened out, has a water-holding capacity of 18–27 times its dry weight.)

The humic colloids in peat soil chelate and also adsorb minerals by cation exchange and release them slowly. Root hairs adsorb nutrients because of colloidal acids in their cell walls, and they also secrete carbon dioxide that goes into solution as carbonic acid and dissociates to provide negative charges for the attraction and concentration of cations. The uptake of anions, of nitrogen, phosphorus, sulfur, chlorine, and boron, is not well understood, but heat energy and oxygen are required. If the air capacity is too low because of compaction or overwatering or if the soil temperature is too low, plants may suffer deficiencies even though anion nutrients are present.

Differences in decomposition and compression make it difficult to estimate the rate of peat accumulation. In the area of the Great Lakes, estimates vary from 100 to 900 years per foot. In a 12-foot deposit in the Boundary Waters region of northern Minnesota, the upper 8 feet took some 7,000 years to accumulate at an average rate of 875 years for each foot, assuming a constant rate of deposition and equal compression (194, 196). But a deposit not far off in northern Wisconsin, 50 feet deep and 9,000 years old, required only 180 years per foot (190). In a southern Michigan deposit, about 18 feet accumulated over 11,000 years, at an apparent rate of 550 years for each foot (286). In the Myrtle Lake region of northern Minnesota, the rate of accumulation was about three times greater during the past 3,000 years than it was earlier. Between 10,316 and 3,160 years ago, only 1.8 inches were deposited in each century, but thereafter the rate was 5.3 inches per century (103).

The world's peat reserves are estimated at 223 billion tons, and more than half of the total is found in the Soviet Union. Large reserves are held in Baltic Europe, the British Isles, the United States, Canada, and Indonesia. About half of the reserves in the United States are located in Alaska. Other states abundant in peat are Minnesota, Michigan, Wisconsin, Maine, New York, North Carolina, Florida, and Louisiana. In the Upper Midwest alone, that is, in Minnesota, Wisconsin, and Michigan, there are close to 15 million acres of peatlands.

Peat Mining

The U.S. peat industry is mainly concerned with highly decomposed reed-sedge peats, or *muck*. Michigan leads in the harvest of such powdery, black "Michigan peat" used as potting soil and lawn dressing. Reed-sedge peat mining is concentrated in the southern half of Michigan's Lower Peninsula. The brown, fibric "Canadian peat" offered on the American market is, for the most part, imported from Canada. It is used mainly as a soil conditioner. Some peat of this type has been harvested in Cheboygan County, in northern Michigan, during the past few years.

Michigan in 1977 produced 268,000 tons of reed-sedge peat; Florida, the nation's second largest producer, dug only 145,000 tons. In 1978 Michigan produced 219,900 short tons of reed-sedge peat, valued at $3,850,000. Illinois, Indiana, and Minnesota together produced only 162,000 tons (136).

Ireland is a country poor in trees. Peat is the only domestic source of fuel. An Irish household may use as much as 15 tons of dry peat fuel each year. A strong man might work two weeks or even more hand-cutting sod for the cooking and heating needs of his household (fig. 7-1). As recently as 1955 about 6,000 men housed in specially built hostels earned their living by cutting Irish sod peat for fuel. But mechanization entirely replaced handcutting by 1960. *Milled peat* (fig. 7-2), shredded, fluffed up for drying, and sometimes compressed into briquettes, is far more economically produced than *block-cut peat* (fig. 7-3), whether cut by hand or by machine. *Machine peat*, produced by a hydraulic method by which peats of all degrees of humification are macerated and combined in a slurry before being pressed into briquettes, makes use of layers that cannot be block-cut, owing to structural weakness, or milled, owing to compaction.

Next to the Soviet Union, Ireland leads the world in fuel peat production (about 4.5 million tons per year). Each year in the late seventies, Ireland used 3 million tons of milled peat for generating electricity, and in addition 0.5 million tons of milled peat briquettes for domestic fuel and industrial heat and steam, and 1 million tons of block-cut peat for domestic fuel. In addition, Ireland exported an annual 1 million cubic meters of horticultural peat moss. During the mid-seventies the Soviet Union burned 70 million tons of peat each year, mainly in generating electricity. Peat fuel is also used in Finland, Sweden, Germany, and Poland but has scarcely been used in the United States and Canada. Following the coal miners' strike in 1902–3, the U.S. Geological Survey inventoried the nation's peat, and a renewed interest in fuel peat resulted from the energy crisis of the seventies.

America's peat resources are estimated to exceed the combined energy potential of the nation's petroleum, lignite, and natural gas reserves. Fuel-grade peat has a high caloric value, more than 8,000 Btu per dry pound. One ton of air-dried peat has the energy equivalent of one-half ton of coal. Peat fuel has a low sulfur content and emits fewer pollutants than coal. Peat can be converted into "synthetic natural gas" as an even cleaner type of fuel. Because of escalating costs of petroleum and natural gas, peat reserves in the United States may indeed assume economic significance in the near future.

It is estimated that Michigan's fuel-grade resources equal 1.2 billion tons, most of it in the Upper Peninsula (where a recent inventory was made by researchers under contract with the U.S. Department of Energy and the Michigan Energy and Resource Research Association). State lands include a considerable acreage in peat. The exploitation of muck and reed-sedge peat has been concentrated in the southern half of the Lower Peninsula, on private land.

Minnesota is much engaged in peatland studies. The Great Lakes Peat Products Company, in Minnesota, is testing compressed peat as an industrial source of energy. The Minnesota Gas Company holds a lease on some 200,000 acres of peatlands with a view toward producing methane gas. In North Carolina, a corporation called the First Colony Farms has been vacuum harvesting peat from 372,000 acres. This peat, largely woody in origin, is deposited to a depth of about 6 feet. Plans are being developed for the production of gas or electricity.

Fig. 7-1. Hand winning peat in Ireland, showing traditional tools used in cutting turf. The man is using a spadelike slane with two cutting edges. (Provided by Bord na Móna.)

Whether a peatland can be profitably exploited for horticultural peat depends on location and climate, as well as size of deposit and physical and chemical nature of the peat. Accessibility is a prime consideration, and crucial to the success of the operation is the cost of transport. Trucking is generally more reliable and economical than rail transport, but 75 percent of the peat sales take place in the spring when competition for trucking is keen. Moderate winds and low precipitation during the harvest season are requisite. In the Canadian Maritime Provinces, as in northern Michigan, about 100 days, from mid-June to late September, can be counted on. Stockpiled peat can sometimes be baled as late as November, although

Fig. 7-2. Peat milling machinery

Fig. 7-3. Bagger, used for block-cutting

Figs. 7-2–3. Modern methods of peat harvesting in Ireland. (Provided by Bord na Móna.)

reabsorption of moisture can be a problem that late. In spite of mechanization, much hand labor is required, and only an area of depressed economy is likely to supply a seasonal labor force.

In the Canadian Maritime, the minimum bog area feasible for exploitation is 200 acres, requiring, as of 1979, an investment of $600,000 Canadian and yielding some 200,000 bales per year (a bale being 6 cubic feet) (187). At Dingman Marsh in Cheboygan County, Michigan, the Black Forest Peat Company estimated that 1,400 acres of fibric/hemic peat, with a mean depth of 6.3 feet, would yield 60–84 million bales, which at $3.50–4.25 per bale would bring in $210–357 million (138). (Plans to harvest this state-owned land, under royalty, were later abandoned.)

Specialized equipment is needed throughout, from ditching to baling (figs. 7-4–9). Drainage ditches are dug one year in advance of clearing (fig. 7-10). In addition to ditching the perimeter, shallower ditches are dug about 60 feet apart, and the surface between them is contoured for better surface drainage. A settling basin reduces the amount of suspended material going into the discharge stream. (Brush should not be burned on the mat, as fire is an ever-present hazard. Fire equipment is essential as a standby at every phase of the harvest operation.) Harrowing fluffs up the peat surface for drying. One harrow (fig. 7-5) can prepare the surface for five vacuum harvesters. Each harvester (figs. 7-4, 7-6–7) can pick up a load in about fifteen minutes. During the season, as harrowing and harvesting continue, a harvester can gather enough peat for 30,000 bales. About 2 inches of peat are harvested each year. Diesel engines heat up less than gasoline types, and dual-wheel tractors with flotation tires disturb the surface less than half-tracks. The stockpiled peat is moved about with front-end loaders for continued drying to about 60 percent water content by weight and also for preventing spontaneous combustion. It is eventually trucked to a processing plant for further drying, vibration screening, and packaging (figs. 7-8–10).

Instead of vacuum harvesting some operators disk the cleared surface several times to break up large clumps of peat and hasten drying, and in some large-scale operations the upper half-inch is milled by large rotating drums with extended spikes. After one or two days of drying, the loosened peat is ridged by windrowing machines and picked up by tractors equipped with a lift scoop or conveyor.

Fig. 7-4. Vacuum harvesters picking up horticultural peat at Lake Sixteen, Cheboygan County, Michigan

Fig. 7-5. Harrow used to fluff up and dry the surface preparatory to vacuum harvesting

Fig. 7-7. Vacuum portion of a harvesting machine

Fig. 7-6. Vacuum harvest machines

Figs. 7-5–7. Peat harvesting machinery in use at Lake Sixteen, Cheboygan County, Michigan

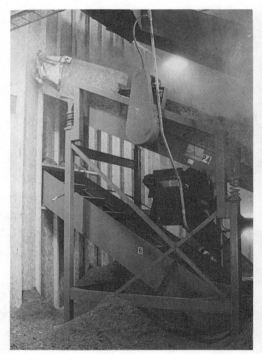 

Fig. 7-8 **Fig. 7-9**

Figs. 7-8–9. Screening and baling horticultural peat, Lake Sixteen, Cheboygan County, Michigan

Fig. 7-10. Lake Sixteen, Cheboygan County, Michigan. Harvested peat stockpiled near the peripheral drainage ditch.

Peat Cropping

> Then plough deep while the sluggard sleeps, and you shall have corn to sell and to keep.
> —Franklin

The success of silviculture on harvested peatlands depends on draining and fertilizing. In order to sustain forest growth in Sweden, the minimum nutrient values in the upper 8 inches of peat need to be 1,600–2,400 lb./acre of nitrogen, 120–160 lb./acre of phosphorus, and 80–100 lb./acre of potassium. In Finland, already forested peatlands make good response to drainage and fertilization. Early thinning, chemical control of competing vegetation, and fertilization increase the yield, and it appears that forest plantations require similar care (187). In Newfoundland, the peatlands are largely treeless, and growth conditions are quite different, owing to high humidity and strong winds. However, experimental and ecological data show that a substantial portion of Newfoundland's peatlands can support merchantable forest stands, especially with improved drainage and fertilization. Fens would, of course, require less fertilizing than bogs. Whether afforestation is economically feasible in Newfoundland is yet to be demonstrated.

The agricultural potential of peat soils is reduced in the north by factors other than the nature of the peat—short summers with long days, for example. Cool climate crops—those requiring a long season for vegetable growth or a second season for flowering and fruiting—do well on peat. They include root crops (carrots, radishes, beets, turnips, rutabagas, potatoes, onions, shallots, leeks, and plants like tulips, grown for bulbs) and leafy or vegetative crops (lettuce, spinach, swiss chard, celery, kale, cabbage, broccoli, cauliflower, brussels sprouts, and mint). Plants grown for fruits and seeds (peas, beans, tomatoes, peppers, egg plants, and corn) do better on clay or loam. Grasses and some legumes, like clover, thrive on peat. Wild rice, cranberries, blueberries, and strawberries can be cropped on peatlands in reclamation. Tree cropping, for fuel, pulp, timber, or Christmas trees, is a good use for mined peatlands, as are crops designed for biomass conversion, for energy, such as willows, alders, hybrid birch, and hybrid aspens. It is possible to produce cattle feed by separating proteinaceous material from the bark of aspen harvested from used peatlands. Fast-growing cattails or aspens can be grown on harvested peatlands and used for producing methane and alcohol.

Recent Estimates of Peatland Crop Production in the Midwest, in Acres (136)

Sod			*Carrots*	
Michigan	10,000		Michigan	6,000
Wisconsin	2,000		Wisconsin,	4,500
Indiana and Ohio	1,000		Minnesota	1,000
Onions			*Radishes*	
Michigan	6,500		Michigan	2,500
Wisconsin	1,500		Ohio	1,800
Celery			*Potatoes*	
Michigan	3,000		Wisconsin	8,000
Lettuce, Endive, Escarole			Michigan	5,000
Michigan	1,200		Indiana	3,500
Wisconsin	1,000		*Mint*	
Cranberries			Wisconsin	15,000
Wisconsin	7,200		Indiana	13,000
			Michigan	6,000

In the Great Lakes area, peat soils are used for hay and pasture. Red fescue, Canada bluegrass, and Kentucky bluegrass do well on peat, but unfortunately much of the acreage is in native grass and poorly drained and unfertilized. However, extensive acreage is devoted to sod farming with Kentucky blue grass.

Cranberry production makes use of peat beds covered with sand. Flooding was once used to prevent frost damage, but sprinkling uses less water and is quicker. The crop is flooded, however, at the time of harvest.

Peat farming presents problems. Draining and clearing are difficult. The soil is deficient in nutrients, wetness promotes some plant diseases, and peat lacks firmness as a rooting medium. Late thawing, frost heaving, and early frost shorten the growth season. Muck absorbs heat by day but loses it quickly at night. Because of heat absorption and poor heat conduction, mucky soils can get too hot for small plants. To avoid plant burn-off, the soil should be kept moist and firm, and windbreaks should be spaced to allow the circulation of air. A considerable problem is peat loss, or "subsidence." Drained and aerated muck (and other peat) rapidly decomposes and when dry blows. In southeastern England, where intensive agriculture has replaced vast areas of wet fen, roads and even drainage ditches may be conspicuously elevated above tilled fields owing to subsidence. At Holme Fen an iron column driven into peat before the beginning of drainage in 1851 was exposed by 1951 to a depth of 11 feet 2 inches (91). Windblown muck can clog drainage ditches, and surface layers of dry muck may resist wetting and cause water stress. To reduce wind loss, tilling and strip-cropping at right angles to the prevailing winds are recommended. Fall-planted cover crops such as rye and buckwheat, snow fences, and windbreaks are also helpful. Peat also burns and produces a great deal of smoke. Fires smoldering beneath the surface, sometimes over winter, are difficult to fight, and what farmer is equipped for fighting fire? Ditching down to a wet level localizes the fire and allows it to burn itself out, but compacting the soil by heavy tractors and rollers is the most effective way to fight muck fires.

Humic acids dry irreversibly. Because of the high content of humic acids in decomposed grades of peat, hard clods difficult to rewet form on drying. Certain wetting agents can be used in the greenhouse. Outdoors, drying is less troublesome because winter frost expansion causes peat clods to break up. The resulting crumb structure provides a greater surface for ion exchange. Liming the soil, allowing it to freeze and thaw, and then ploughing help provide a good structure. For greenhouse culture a 3:1 mixture of sand and peat gives anchorage for plants that otherwise require support. Plants such as tomatoes grown under glass on rich loams may be troubled by a buildup of toxins or disease organisms. A *Sphagnum* peat mixture giving improved aeration and drainage (and perhaps an antibiotic effect) offers a simple solution to such problems.

The roots of most plants will not penetrate water-saturated soil. Oxygen, which helps to convert insoluble nutrients to a usable form, is essential for root hair development and seed germination. Some plants of wet, even aquatic habitats can stand long periods of anoxia because of aerating tissues in stems and roots, but others cannot. Even cranberries which live in wet places and remain dormant for long periods of time cannot, and the trees of bogs, including black spruce, will die if the water level is raised over a prolonged period. The effect of flooding on the gas balance at times of peak biotic activity, microbial or otherwise, is particularly damaging. Under conditions of oxygen shortage, the absorption of nutrients is adversely affected in the order of potassium, calcium, magnesium, nitrogen, and finally phosphorus. The supply of reduced (ferrous) iron is higher under acid conditions of oxygen deficiency, and high levels of reduced iron—and manganese too—can injure roots. Also, hydrogen sulfide reaches toxic levels of concentration, and denitrification increases.

Sedge peats and mucks derived from them are richer in minerals than *Sphagnum* peats, yet they too need to be limed and fertilized, copiously, with nitrogen, phosphorus, and potassium. In coastal Newfoundland, where scarcely any mineral soil can be put to cultivation, composting peat with fish and kelp provides economy in fertilizing. Fertilizing peat soils creates a microbial bloom that results in a greater release of nutrients.

The power of roots to accumulate nutrients is directly related to temperature. Low temperatures lower the rate of nitrogen assimilation. They also reduce water absorption to the point of wilting even under conditions of saturation. This reduction is related to retardation of root tip elongation and root hair formation and also to decreased permeability of cell membranes. When low temperatures accompany poor aeration, roots may lack the vigor needed to resist infection.

A good horticultural peat has a fibrous structure and an absorptive capacity greater than twelve times its weight, an ash content less than 5 percent, and a pH of 3.5–6. *Sphagnum* moss, only slightly decomposed, lightens soils, allowing air and water to move freely. Aeration increases microbial activity and mineral release. *Sphagnum* increases the water-holding capacity of the soil and prevents excess drying. It takes up and holds minerals and improves the buffering capacity of the soil. Sedge peats are prone to drying and quickly lost by decomposition, but *Sphagnum* peat, fibric in texture, lasts for years.

Consumption of horticultural peat in North America is considerably less than in Europe because of competitive substitutes—straw, peanut hulls, sawdust, and bark, not to mention inorganic additives like vermiculite, perlite, and pumice. *Sphagnum* mulch is commonly used for such heath plants as rhododendrons. Such acidophiles can be planted in large blocks of peat piled up in rock-garden fashion or moved about on castors. Powdered peat is mixed with fertilizer to reduce caking before and after application and also processed into containers for growing plants. *Sphagnum* combined with fertilizer and a wetting agent is pressed into disks and squares that swell up on wetting to form cylinders and cubes used in germinating seeds and rooting cuttings. "Containerized" seedlings are much used in the forest industry.

Health Hazards Associated with Peat

Persons handling horticultural *Sphagnum* run some risk of contracting sporotrichosis. The disease, caused by the fungus *Sporothrix schenckii*, enters the body through cuts and scratches. In one to four weeks, small, painless blisters appear. They become inflamed and slowly enlarge. If untreated, the disease slowly progresses to other parts of the skin and, through the lymphatic system, to bones, lungs, and abdominal organs. Scarring is common. The disease is not contagious. It is chronic but not often fatal. Delayed diagnosis is a problem because few physicians know the disease. The infection does not respond to antibiotics. Dosage with potassium iodide, given orally over several months, is usually effective. Persons making horticultural use of *Sphagnum* are advised to wash their hands frequently, give prompt attention to cuts and scrapes, even wear rubber gloves. (The fungus, also known as *Sporotrichum schenckii*, is not limited to peat. It occurs in soil and on vegetation, and infections most often result from scratches by thorns and splinters.)

Probably the greatest hazard to peatland farmers and peat harvesters is breathing muck dust. In the Kankakee River valley, in northern Indiana and Illinois, farmers may be plagued by "itchy muck" because of freshwater sponge spicules contained in blowing sandy muck.

The Varied Other Uses of Sphagnum and Peat

> Be not simply good, be good for something.
> —Thoreau

In desperation one might use anything, and so it is that many of the uses for *Sphagnum* and peat recorded in the literature are more quaint and curious than significant. One reads that Laplanders put *Sphag-*

num in bread and the Chinese use it as starvation food. Alaskan Indians treat cuts with a salve made of *Sphagnum* leaves and grease, and the Chinese boil *Sphagnum* in water as a cure for hemorrhages and eye diseases. Actually *Sphagnum* provides an antibiotic action against both gram negative and gram positive microorganisms. Peat tar is said to possess antiseptic and preservative properties, and sphagnol, a ligninlike distillate of peat tar, is "authoritatively recognized" as a treatment for various skin diseases and relieves the itch of mosquito bites.

The ability of *Sphagnum* to soak up and retain liquids has proved useful in staunching blood. *Sphagnum* takes up considerably more moisture than cotton. In cotton, water is simply enmeshed in a tangle of fibers, but *Sphagnum* holds water in and among its stems and leaves owing to a spongelike construction, even down to the cellular level. The absorbent nature of *Sphagnum* is also related to a scant amount of lipids produced by the plants. North Europeans have, for centuries, used *Sphagnum* in dressing boils and discharging wounds of man and his beasts. *Sphagnum* was used to some extent as a surgical dressing during the Napoleonic Wars and recommended for use, at least, during the Franco-Prussian War. The Japanese found use for it during the Russo-Japanese War. During the First World War such absorbent pads had a special appeal because *Sphagnum* replaced hard-to-get cotton and made more of it available for use as gun wadding (158, 161). In 1918 the British used 1,000,000 *Sphagnum* dressings each month. In that year Scotland was asked to supply 4,000,000 pads per month and Canada 20,000,000. Canada actually produced about 200,000 per month, and from March to September of 1918, 500,000 dressings were prepared in the United States. The pads, covered with cotton gauze, were considered superior to all-cotton dressings. They absorbed liquids three times as fast, took up three to four times as much, and distributed them better. They were cooler, less irritating, and much cheaper.

Indians have used *Sphagnum* in lieu of menstrual pads and diapers. (It appears to prevent or clear up diaper rash.) Eskimos use *Sphagnum* as boot liners. It is possible to buy compacted *Sphagnum* insoles, said to absorb smells as well as moisture while providing a cushioned comfort. Peat can be used to absorb oil spills on water. It makes an absorbent litter for cattle barns.

Scotch whisky (and Irish whisky too) is produced from germinated barley, or malt, dried over open peat fires. The smoky flavor varies also with the quality of the water, commonly trickled through Highland peat bogs, and also with the casks used for aging, sometimes barrels previously used for sherry wine.

Sphagnum has been used for lamp wicks and mattresses, steeped in tar and used for caulking watercraft, and stuffed between timbers to insulate houses and deaden sound. Mixed with molasses, it has been fed to livestock. It is burned to smudge crops in danger of freezing. Woven in with wool, it has gone into textiles. Peat treated with alkali produces a brown dye. It can be incorporated into wrapping paper, wallpaper, building paper, and wallboard. It can be mixed with clay to make porous bricks or Portland cement to produce lightweight "peatcrete" blocks that can be sawed and fitted. It can be pressed into a form simulating lumber.

Uses of some commercial importance or potential include the extraction of mineral and organic substances. Peat products have included paraffin, naphtha, ammonium sulfate, acetic acid, ethyl and methyl alcohol, and illuminating gas. Phenols, waxes, and resins can be extracted from peat. The hemicelluloses, celluloses, and polypeptides contained in peat can be converted into materials used in producing a yeast protein for cattle feed. In the Soviet Union one million tons of single-cell protein are produced in this way each year. Humic acids are used in China and the Soviet Union as viscosity-modifying agents in oil drilling muds and also as additives to decrease the cost of cement structures. Peat converted to coke can be used in processing iron ore. It can be made into activated charcoal for purifying and decolorizing water and lignitized to provide a coal substitute. Peat is also used to bind and remove heavy metals from industrial effluents.

Currently under investigation is the use of wetlands in waste water treatment, sludge composting, and nutrient cycling (277). Wetlands are cheap, and they can be used economically in removing nitrogen, phosphorus, and other materials from sewage. Peat soil and the plants rooted in it can remove 96 percent of the nitrogen and 97 percent of the phosphorus applied as eutrophic river water. Muck soils can adsorb 6–12 percent of the nitrogen and 60 percent of the phosphorus applied as sewage. Grasses on loam remove 42 percent of the nitrogen and 30 percent of the phosphorus from sewage, but grasses on muck remove as much as 70 percent of the applied phosphorus. Significant reductions in phosphorus, nitrogen, calcium, magnesium, potassium, and trace elements can be attributed to cation and anion exchange mechanisms in the soil, as well as plant assimilation and, especially, microbial decomposition. Grass crops are especially effective in removing nitrogen and phosphorus because the penetration of fibrous root systems aids in infiltration and aeration. Although mineral soils are also efficient in cleansing sewage effluents, long-term use results in nitrogen pollution, but poor drainage or short periods of waterlogging of wetland soils promote denitrification. The long-term efficiency of using wetlands in sewage disposal is not known, but the method seems suited to small towns and campgrounds where more costly systems are not feasible.

Peat Conservation

> Enjoy the land, but own it not.
>
> —Thoreau

"Peat harvesting" is no more than a euphemism for mining. The British "peat winning" has more pleasant connotations, but harvesting, however inaccurate, is commonly used because it has the ring of familiarity. It should be understood, however, that the "harvesting" can do irreparable harm to the ecosystem. Ombrotrophic peatlands in Finland have accumulated peat so slowly over the past millennium or two—only 4–16 inches per thousand years—that regeneration would seem to exist only in terms of geological time (253). The blanket bogs of oceanic Great Britain have slowly regenerated from ancient digging and prolonged grazing, but it is highly unlikely that they can survive and recover from recent large-scale onslaughts. It is unlikely that regeneration is possible in the Great Lakes region. It has been observed, in the more favorable climates of Maine, that undrained peatlands cleared of vegetation regenerate within a few years, but deep-ditched peatlands will not recover even decades after drainage. *Sphagnum* is virtually lacking, shrub growth sparse, and species diversity low. Because *Sphagnum* will not grow where the water table is lowered, owing to difficulties in getting capillary water, the early stages of succession cannot be repeated. Drained mires, even if vegetated, may not accumulate additional peat and may, in fact, lose peat by degradation.

In North America the peat resources have not yet been exploited to the extent that irreparable damage to the peatland ecosystem is of timely concern. However, the history of depredation in Great Britain, especially in recent times, is worthy of consideration. It can happen here!

The cool, moist, westerly winds off the Atlantic Ocean carry with them enough moisture to produce measurable precipitation on two out of every three days over parts of Wales, England, Scotland, and Ireland. The Gulf Stream moderates the climate, maintaining an even coolness throughout the year and ensuring that most of the precipitation falls as rain rather than snow. This climate produces vast, rolling, treeless moors, or blanket bogs, in the most oceanic parts of the British Isles. Blanket bogs are also found in western Norway and some more distant parts of the world, but they are best developed in the British Isles. The blanket bog ecosystem is well worth protecting, yet Britain's peatlands are being eroded away at an alarming rate owing to increased use in peat cutting, agriculture, and forestry. Between 1850 and 1978, 87 percent of Britain's lowland raised bogs were destroyed (133, 151, 249). Nearly all of the great raised bogs of central Ireland have, in recent decades, been stripped away by the Bord na Móna, or national peat board, for peat used to fuel power stations. Many of Britain's upland blanket bogs have been drained and planted to forests. The "Great Flow" that spans the breadth of northern Scotland is now being transformed into forest plantations at the rate of 140 acres per week.

Does the large-scale use of peatlands matter? Those who favor transforming blanket bogs into forests argue that they are making use of unproductive wilderness and thereby providing employment opportunities as well as a long-term source of revenue. Conservationists, on the other hand, argue that the quality of timber grown in the Scottish Highlands scarcely justifies the high environmental cost. The habitat is marginal for most trees, and the lodgepole pines and Sitka spruce planted there grow very slowly and readily blow over. They argue that the blanket bogs, or flows, that took thousands of years to form are being destroyed forever. The gross transformation of a unique ecosystem into poor quality forest adversely affects the wildlife of the open moors. For example, forest cover favors predators that endanger birds nesting nearby in the open. Grouse are endangered by alterations in the moorland habitat, and salmon are adversely affected by silting and chemical alterations in streams. The threat to grouse shooting and salmon fishing is an important consideration in an area where controlled hunting and fishing provide a major source of income from tourism.

As a natural resource, peat should indeed be used, but peatlands have some values greater than monetary. Environmental concerns include aesthetic diversity, protection of unusual ecosystems and the plants and animals characterizing them, and watershed management. Peatlands should be exploited only with some reasonable plan, such as sequential logging, harvesting, and cropping. There is no hope of returning mined peatlands to their former state, but they can be continued in usefulness.

Peatland exploration should, by all means, be followed up by peatland reclamation. European countries sometimes require developers to leave one or two feet of peat above the mineral soil or a mixture of the two as an agricultural soil. Contouring for water control may also be required. The Soviet Union and Scotland have been successful in agricultural use of harvested peatlands. Ireland expects to use up the large peat deposits of the Midlands by the end of the century. Meanwhile, the Bord na Móna is finding ways to reclaim the land, for afforestation, agriculture, and horticulture, as rhododendron nurseries, for example. In Finland, reclaimed peatlands are, for the most part, put into Scotch pine, red pine, spruce, aspen, and birch. Canada is experimenting with silviculture and also with impoundments created in excavated peatlands as habitats for waterfowl and fur-bearing animals. (One might wonder whether such wildlife usage is economically best for a country of unlimited wetness.) Newfoundland, Canada's easternmost province, has made great progress toward understanding the agricultural potential of peatland, whether exploited or not.

Peat on public lands should be exploited only under control, by permit and contractual agreement safeguarding biological and aesthetic values and guaranteeing reclamation. (And reclamation should be effectively carried out. Wild rice unharvested or blueberries left untended serve no useful purpose.) Environmental impact studies during and after exploitation should include monitoring the water level and quality in adjacent peatlands, as well as measuring the rate of outflow and its effect on silting and stream bank erosion below the area of discharge. The government of Newfoundland curbs the uncontrolled use of peat deposits by requiring a royalty payment of fifteen cents per cubic meter. Newfoundland needs a fuel source, and the peatlands there have an enormous potential to meet energy and agricultural needs. The government has therefore taken the initiative in exploring that potential. In Michigan, the need has not been great, certainly not for fuel peat, and the peat resources have been inventoried only in part. Peat deposits are subject to state control only under laws regarding mineral rights, and the Department of Natural Resources has neither the knowledge nor the personnel to plan for long-range controls. Only recently was application first made for peat harvesting on state land, at Dingman's Marsh in Cheboygan County. Shortly before that request was made, by the Black Forest Peat Company, the state arranged a trade, with the same company, whereby part of a large and beautiful bog mat at Lake Sixteen, also in Cheboygan County, and a marshy site occupied by the rare *Juncus militaris* were exchanged for adjacent state lands. The company began peat harvesting, on its own land, under an agreement to monitor ecological impacts on the state's peatland. (The company went out of business in 1987. Dingman's Marsh was not exploited, although a permit was given, with provision made for royalty payments to the state.)

Chapter 8

A Close Look at Peat Mosses: A Bryologist's Vademecum

> The universe is wider than our views of it.
>
> —Thoreau

Sphagnum is aptly known as peat moss, as peat-producing wetlands are commonly dominated by that genus. It takes no training at all to recognize the genus and note its abundant beauty in a peatland. The species can be almost as easy to recognize. They flourish in acid bogs and also in open fens and coniferous swamps where conditions are base rich. *Sphagnum* species initiate and control many of the vegetational changes in peatland development, and they are effective indicators of the environmental conditions associated with such changes. Yet, they are scarcely noted by field botanists, and even bryologists avoid them, without justification, as taxonomically troublesome. The species are knowable. They can be sorted out by ecological probabilities and by size, color, and form, as well as microscopic differences.

On a local basis, at least, the species of *Sphagnum* are rather few. In the Great Lakes region there are fewer than thirty species. No more than two or three can be expected as pioneers at the edge of floating mats. Four to six can be expected in grounded mats occupied by a hummock-hollow complex, and each of them grows at a level suited to its moisture and pH preferences. Five species characterize the mineral-rich white cedar swamps, four of them virtually limited to that habitat. Certain species can be found in marly rich fens or alder swamps irrigated by groundwater. Some grow submerged in shallow water. Some grow in compact cushions, others in loose, extensive lawns.

Recognition can be direct and easy. For example, in open peatlands, a stout, red *Sphagnum* with fat branches can be none other than *S. magellanicum*. *Sphagnum papillosum* is similar but golden-brown. It is limited to more mineral-rich, pioneering situations. A similar though pale-green species typical of *Thuja* swamps is *S. centrale*. There is normally no need to seek out the microscopic features that characterize these species. But *S. magellanicum* growing, atypically, in the shade can be green and therefore indistinguishable in aspect from *S. centrale*. *Sphagnum palustre* tends to be yellowish and can be confounded with *S. papillosum* or *S. centrale* in more southerly parts of the Great Lakes region where it is common in mineral-rich, shrubby habitats. In general, these and other species are recognizable in the field and in an altered, though characteristic, state as dried specimens. But some few look-alikes require microscopic examination.

Life History and Structure

It expedites identification, both in the field and in the laboratory, to recognize the sections of the genus at sight and to know the suite of microscopic characters associated with them. Many key characters are sectional in nature, and it helps to skip over them and trouble with only a few technicalities of specific importance. To make a start toward naming sections and species with precision, one needs a basic understanding of the life history and structure of the peat mosses.

The single-celled *spore* (figs. 8-1, 8-3) is the first cell of the haploid gametophytic generation. It gives rise on germination to a juvenile stage called the *protonema* (figs. 8-2, 8-5–6). Normally a small, lobed thallus, the protonema may in poorly lit, liquid culture be branched-filamentous. One or sometimes two leafy shoots are produced at the thallus margin, but marginal filaments can proliferate secondary protonemata at their tips. Even though spores are produced in great abundance and germination is easily induced in culture, juvenile stages of growth are rarely seen in nature. (But Cyrus McQueen succeeded in finding protonemata, in Vermont, by germinating spores in the laboratory in late August and on first seeing gametophores some five weeks later, looking for protonemata sprouting gametophores in the field. Thus, October is a good time to look for protonemata.) Fragmentation and regeneration are also involved in shoot formation.

Rhizoids are restricted to the protonema and the base of young leafy plants, or gametophores. They are filamentous, with slanted crosswalls.

The leafy plants, or *gametophores*, grow in length because of continued divisions of an apical cell with three cutting faces. The stem, commonly forked, is provided with spirally arranged leaves and *fascicles* of branches. The younger fascicles are crowded at the apex of the stem in a headlike tuft, or *capitulum*. A fascicle nearly always consists of relatively stout *spreading branches*, in addition to slender *pendent branches* usually closely investing the stem. The stem consists of a central *parenchyma* of pale, isodiametric cells, a wood cylinder, or *sclerenchyma*, of long, thick-walled, colored cells, and a *cortex* of large, empty, thin-walled hyaline cells. The outer cells of the cortex may have one or more surface pores and, in some cases, reinforcing spiral fibrils as well (figs. 8-7–10), or both fibrils and pores may be lacking.

Stem leaves (figs. 8-12–17) differ from branch leaves in size, shape, and structure. They may be fringed at the margins or bordered by linear cells. The border may be broadened toward the base, sometimes abruptly so. The leaves are made up of a network of narrow cells enclosing large, broadly rhomboidal hyaline cells. The latter commonly lack wall material on one or both surfaces. The *"resorption"* of wall material accounts for fringing at leaf margins (figs. 8-12–15), irregular perforations of walls as *membrane gaps* (fig. 8-29), or wrinkling of thin walls as *membrane pleats* (fig. 8-18). (Resorption refers to the absence and presumably disappearance of wall material, but the process, digestive or otherwise, is not understood.) The hyaline cells are sometimes divided or subdivided (fig. 8-19). In some species, in a normal condition, the stem leaves have hyaline cells reinforced by fibrils, as well as perforations of regular size and shape, called *pores*, and thus approach the branch leaves in structure. Occasional forms of other species may show, atypically, some degree of *isophylly*, presumably in response to unusual conditions of the habitat.

The branch axis consists of a core of parenchyma enveloped by a colored sclerenchyma and a hyaline cortex of one cell layer. The cells of the cortex may have pores and fibrils, but more commonly fibrils are absent and pores are restricted to certain *retort cells* (fig. 8-11) that are enlarged and porose at the end of a protruding neck. Some few species have all the cells of the cortex apically porose and retortlike.

The branch leaves are arranged in such a way that each leaf occupies two-fifths of the circumference of the branch axis and five leaves occupy two complete spirals. This means that every sixth leaf begins a new spiral cycle in line with the first. The branch leaves, commonly larger than those of the stem, are deeply concave and sometimes hooded at the apex. They develop from an apical cell with two cutting faces. As a result of two unequal divisions, each cell derivative gives rise to triads of cells that fit together in a netted fashion, with linear *green cells* enclosing large, empty, rhomboidal hyaline cells strengthened by annular (or rarely spiral) *fibrils* and perforated by pores (figs. 8-21–26, 8-28, 8-30–35, 8-37–39) and only rarely in *gaps* (figs. 8-29, 8-40). The pores differ in number, size, shape, and arrangement on either surface. In some cases connecting fibrils and unperforated *pseudopores* are also present. The pores may have thin or ringed (fig. 8-37) margins. They may occur over the surface in one, two, or more rows, at ends and corners, or along the *commissures*, that is to say, at the junctions of hyaline and green cells (fig. 8-39). The inside surfaces where the hyaline cells join the green cells may be smooth or variously ornamented by fine papillae, a network of fine ridges, or coarse and irregular comb fibrils (figs. 8-36, 8-41–43, 8-49). The green cells, as viewed in section (figs. 8-27, 8-47–49), may be central and entirely included by the adjacent hyaline cells, equally exposed on both surfaces, or triangular to trapezoidal and exposed exclusively or more broadly on one surface or the other. The leaves are, for the most part, bordered by two or more rows of linear cells (figs. 8-44–45) and only rarely toothed except across a truncate tip. In some species a denticulate border results from the resorption of the outer portion of a single row of marginal cells, forming a *resorption furrow* (figs. 8-46, 8-50–51).

The sexual condition may be dioicous or monoicous, varying in some cases between populations of the same species. The male and female sex organs never occur in the same inflorescence, and there are no paraphyses, or sterile threads, mingled with the sex organs, as in true mosses. The *antheridia* (fig. 8-52), globose and long-stalked, are borne singly at the side of leaves crowded together in swollen, catkinlike portions of spreading branches in or below the capitula. The leaves of the catkin are often conspicuously red, orange, or brown and distinctly set in rows. The antheridia may be fully developed in late summer, but the sperms are not ready for discharge until after freezing (fig. 8-53). Plants brought in after freeze-up, in November, for example, may discharge sperm after a few moments of soaking.

Spring thaws, in March through late April, trigger the discharge at a time when water movement can serve in the transport of sperm. (In Vermont, Cyrus McQueen has observed that many of the *Acutifolia* have very young sporophytes already present at first snowmelt in April. Fertilization may have taken place during the preceding fall or perhaps at the time of the January thaw. In Ann Arbor, Michigan, snow does not usually accumulate before Christmas, but there is often a period of unseasonable warmth about December 20. That too may be a time when fertilization of *Sphagnum* can take place.) The antheridia dehisce by means of four broad valves rolling downward from the apex. The *biflagellate sperm* are discharged in vast numbers, though singly rather than in a cigar-shaped cloud, as in true mosses. They gyrate in place and scarcely move any distance at all unless a receptive archegonium is at hand.

The flask-shaped *archegonia* (fig. 8-55), produced in the capitulum later than the antheridia, perhaps in late summer or autumn, occur singly or in groups of two to five, enclosed in large and otherwise differentiated *perichaetial* leaves. A single *egg* is produced in the *venter*, or enlarged base, of each archegonium. The fertilized egg migrates below the venter where a *calyptra* develops as a protective membrane. The calyptra closely invests the developing sporophyte and becomes ruptured only at its maturity. The sporophyte, diploid in chromosome number, consists of a globose *capsule* and a massive *foot* embedded in gametophytic tissue. The sporophyte is elevated at maturity by an elongated gametophytic stalk, or *pseudopodium* (figs. 8-54–55).

The capsule wall consists of several layers of cells in a solid tissue. It has on the surface layer a large number of *pseudostomata* consisting of two somewhat sunken guard cells with no opening between them. Within the capsule, two layers of spore-mother cells surround and overarch a massive, domelike *columella*. (Except in the area of overarch the spore-bearing tissue is amphithecial in origin; that is to say, it develops from outer tissue delimited by the first tangential divisions of the embryo.) Each spore, produced as a result of meiotic divisions and thus haploid in chromosome number, bears a Y-shaped scar as evidence of a tetrad origin (fig. 8-1). The *spores* are large (18–42 µm), tetrahedral, and smooth or variously roughened. They are explosively discharged owing to a buildup of air pressure as the capsule dries up and shrinks until the flat lid, or *operculum*, pops off.

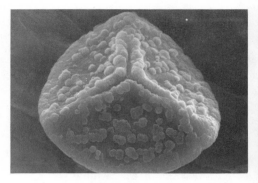

Fig. 8-1. Spore of *Sphagnum capillifolium* var. *tenellum*. (SEM photo by Cyrus McQueen.)

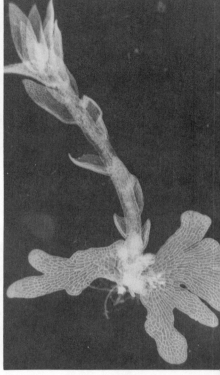

Fig. 8-2. Protonema and young gametophore of *S. junghuhnianum* var. *pseudomolle*. (Photo by Y. Nishida, communicated by A. Noguchi.)

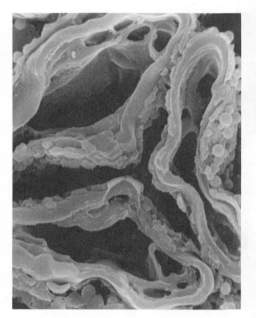

Fig. 8-3. Tetrad of spores in section. (SEM photo by Robert Chau.)

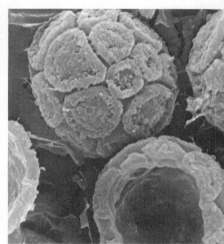

Fig. 8-4. Spores of *Tilletia sphagni*. (SEM photo by Robert Chau.)

Fig. 8-5. A young protonema

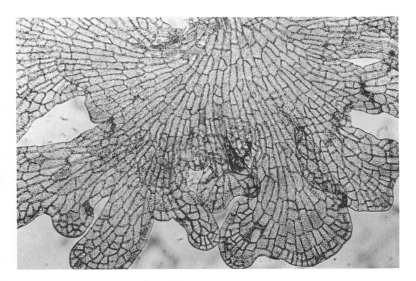

Fig. 8-6. The margin of an older protonema

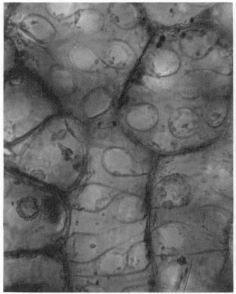

Fig. 8-7. *Sphagnum portoricense*

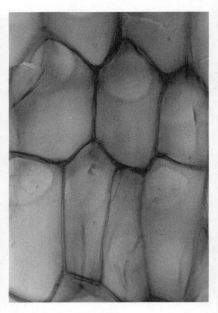

Fig. 8-8. *S. russowii*

Figs. 8-7–9. Epidermal stem cells

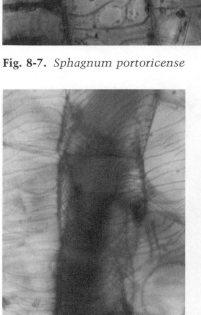

Fig. 8-9. *S. centrale*

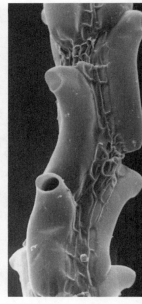

Fig. 8-10. Cells of the branch cortex of *S. portoricense*

Fig. 8-11. Retort cells of the branch cortex of *S. tenellum.* (SEM photo by Jeffrey Holcombe.)

 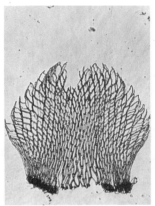 

Fig. 8-12. *Sphagnum riparium*

Fig. 8-13. *S. fimbriatum*

Fig. 8-14. *S. girgensohnii*

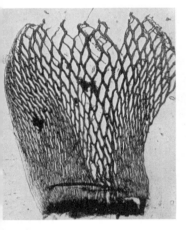  

Fig. 8-15. *S. lindbergii*

Fig. 8-16. *S. teres*

Fig. 8-17. *S. capillifolium* var. *tenellum*. (Photo by Dale Vitt.)

Figs. 8-12–17. Stem leaves

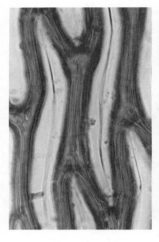
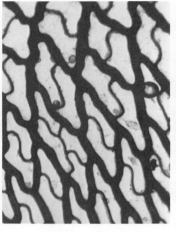
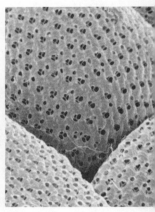

Fig. 8-18. Stem leaf cells of *Sphagnum russowii*, showing membrane pleats

Fig. 8-19. Divided stem leaf cells of *S. papillosum*

Fig. 8-20. *S. magellanicum* outer surface of branch leaf showing pores grouped in at adjacent cell angles. (SEM photo by David Levick.)

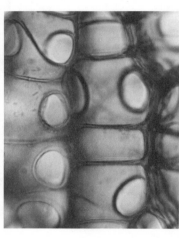
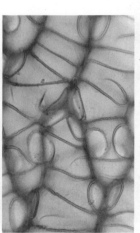
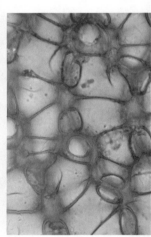

Fig. 8-21

Fig. 8-22

Fig. 8-23. *S. portoricense*

Figs. 8-21–22. *Sphagnum henryense*, upper cells, outer surface

Figs. 8-21–29. Structural details of branch leaves

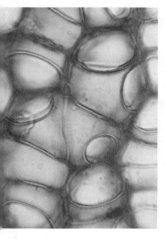 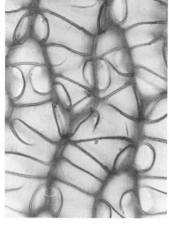

Figs. 8-24–25. *Sphagnum palustre*

8-24   Fig. 8-25

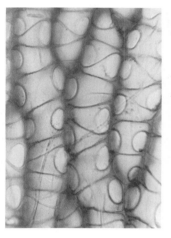 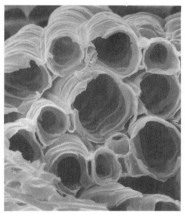

Fig. 8-26. *S. henryense*, lower cells, outer surface

Fig. 8-27. *S. obtusum*, leaves in section. (SEM photo by Jeffrey Holcombe.)

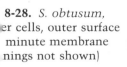

8-28. *S. obtusum*,
er cells, outer surface
minute membrane
nings not shown)

8-29. *S. imbricatum*,
wing membrane gaps
e leaf tip. (SEM
to by Jeffrey
combe.)

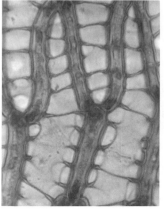 

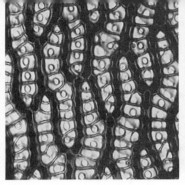 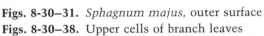 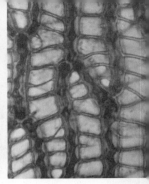

Fig. 8-30

Fig. 8-31

Fig. 8-32. *S. cuspidatum,* outer surface

Figs. 8-30–31. *Sphagnum majus,* outer surface
Figs. 8-30–38. Upper cells of branch leaves

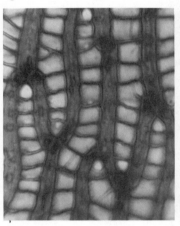 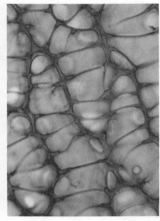 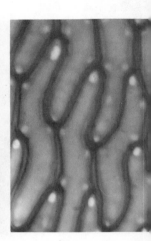

Fig. 8-33. *S. obtusum,* outer surface (showing "window pores" at the ends)

Fig. 8-34. *S. recurvum,* inner surface

Fig. 8-35. *S. splendens,* inner surface. (Photo by Robert Gauthier.)

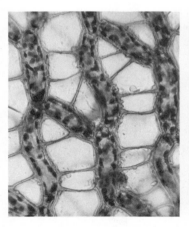 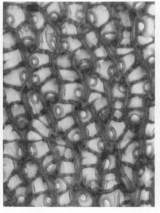 

Fig. 8-36. *S. portoricense,* showing fringe fibrils of side walls of hyaline cells

Fig. 8-37. *S. warnstorfii,* outer surface. (Photo by Jeffrey Holcombe.)

Fig. 8-38. *S. molle,* outer surface, showing bulging hyaline cells and narrowly elliptic commissural pore

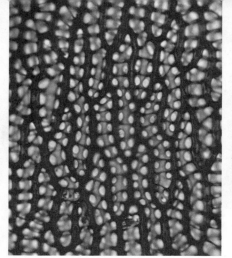

Fig. 8-39. *Sphagnum subsecundum.* Upper median cells of branch leaf, outer surface, showing crowded commissural pores. (Photo by Bodil Lange.)

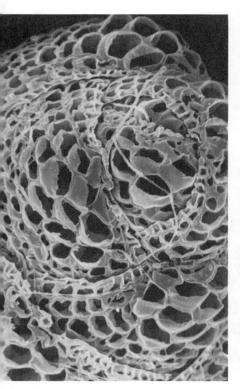

Fig. 8-40. *S. magellanicum*, tip of young branch showing membrane gaps at leaf tips. (SEM photo by David Levick.)

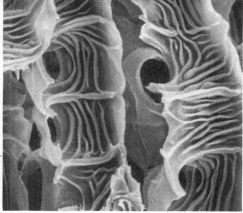

Fig. 8-41. *S. imbricatum*

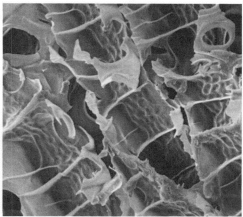

Fig. 8-42. *S. henryense*

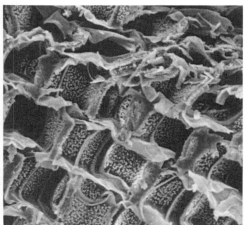

Fig. 8-43. *S. papillosum*

Figs. 8-41–43. Ornamentation of side walls of hyaline cells of branch leaves

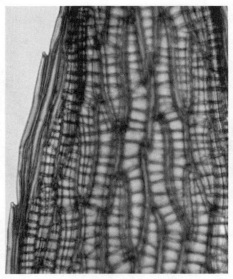

Fig. 8-44. Linear cells of the branch leaf border in *Sphagnum cuspidatum* var. *serrulatum*

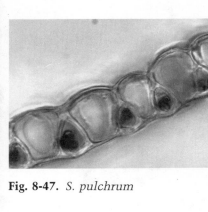

Fig. 8-47. *S. pulchrum*

Fig. 8-45. Linear-celled border of branch leaf in section

Fig. 8-48. *S. centrale*

Fig. 8-46. Resorption furrow of branch leaf in section

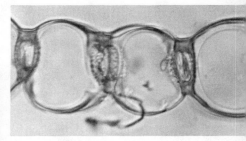

Fig. 8-49. *S. papillosum*. (Photo by Dale Vitt.)

Figs. 8-47–49. Cross-sections of branch le

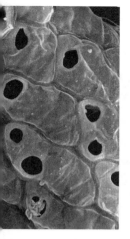

Fig. 8-50. *Sphagnum compactum* **Fig. 8-51.** *S. portoricense*

Figs. 8-50–51. Resorption furrows in surface view. (SEM photos by David Levick.)

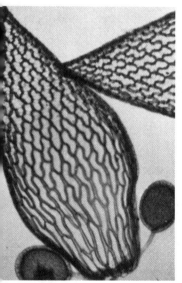

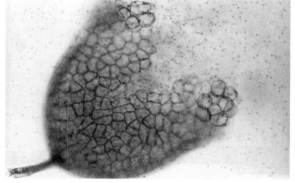

Fig. 8-53. Antheridial dehiscence

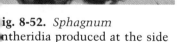

Fig. 8-52. *Sphagnum* antheridia produced at the side of highly colored leaves of a male catkin. (Photo by Jerry Snider.)

Fig. 8-54. Capsules uplifted on pseudopodia. (Photo by E. S. Hains.)

Fig. 8-55. Capsule and pseudopodium with archegonia at the junction between them. (Photo by L. E. Anderson.)

Fig. 8-54

Fig. 8-55

The pressure inside the capsule is said to equal four to six atmospheres—the same pressure carried in the tires of medium-sized truck trailers. The spores, some 16,000–25,000 per capsule, are suddenly hurled more or less in mass to a distance of 3 to 10 centimeters. Dispersal to a nearby wet substrate increases the likelihood of germination in a favorable habitat over that resulting from haphazard, windblown dispersal.

The time between fertilization and spore maturation is about four months. Capsules are commonly produced by all the species and usually in abundance. Spore dispersals take place in the Great Lakes region from late June to mid-August.

The haploid chromosome number is 19, although many species exist in populations with 19 or 38 chromosomes. Whether the difference is related to the variable sexuality that some species exhibit is not known.

Fungal hyphae are commonly seen in the perichaetia, and mature capsules are often filled with small, spherical, reticulate-marked spores of the fungal parasite *Tilletia sphagni* (fig. 8-4).

Methods of Study

Especially at the beginning, it is advisable to coordinate field observation with microscopic study. It is better to collect thoughtfully, first exploring the habitat variation in a peatland and then looking for noticeable differences in aspect in those habitats, than to collect rashly and repetitively, ending up with much duplication and much more material than one can examine carefully and promptly in the laboratory.

A 10–14× handlens is desirable for field study. In the laboratory, a binocular dissecting microscope with magnifications of 10–30× is a great convenience, and a compound microscope is essential, as most characters require high-dry magnifications of about 420×. Dry material has its own characteristic look, depending on the species, and it can be soaked up quickly in hot water or water to which a little detergent or alcohol has been added. Sharp-pointed forceps, a spearpoint needle, a few single-edge razor blades, a stick of elder pith, slides, coverslips, a couple of dropping bottles, perhaps a watch glass, and an old tea towel are sufficient.

Each person needs to develop his own techniques, yet a few suggestions may help. In order to observe the structural details of stem cortex and stem leaves, the capitulum and branch fascicles should be plucked off. The stem should be stained and washed before scraping. For branch leaves and branch cortex, a well-developed spreading branch should be stained before scraping from the fatter, lower portions. For sectioning branch leaves, a quick-slice method gives good results. A wet branch, unstained, can be placed on a slide under a dissecting microscope and held down with another slide or a small piece of pith, flat at the bottom and beveled at the end. While resting a razor blade against one's fingernail, many slices can be made under control. Many of the sections will be, if not perfect, good enough! (It is also possible to place the material in a fold of wet paper and slice through both.)

For sections of stems or branch leaves staining is undesirable, but for other structures essential. A quick method is to puddle the tip of an indelible copying pencil in a water mount before adding a coverslip. (The same pencil can be wetted in the field to show stem leaf characters. A felt-tipped pen can also be used.) Better contrast can be achieved by using a prepared stain and washing and blotting away the excess before dissection. Staining can be done by dipping in the solution or adding drops to the material on a watchglass, slide, or paper towel. Crystal violet in a saturated solution with water or 50 percent ethyl alcohol gives a strong purple in a few moments. A pleasantly blue stain of lesser contrast is given by methylene blue mixed in a 1–2 percent solution in water or 50 percent alcohol. (The alcohol expedites soaking and staining. A few drops of glycerol added to a dropping bottle of stain stabilize alcohol solutions.)

Reference slides have only limited value. The species are relatively few and their diagnostic features easily remembered. It is easy enough to make comparisons with slides newly prepared from voucher specimens or with appropriate illustrations. If a permanent mount is desired, however, some medium other than Hoyer's solution is needed. Hoyer's, so useful for dissections of true mosses, leaches out the stain used on *Sphagnum* preparations. The following mounting medium has been used:

Make preparations as usual in water, including staining, then add a few drops of the mounting medium (1:1 solution of water and corn syrup plus a few drops of liquid phenol, 2 percent). Add a little more stain before applying a coverslip. Ring the coverslip with nail polish or aluminum paint.

A collection of dried voucher specimens is recommended for purposes of comparison and also for their scientific value as distributional records. *Sphagna* do not need to be dried quickly and should not be pressed tightly. Flat clumps about the size of the palm of one's hand can be laid out in newsprint and stacked for drying. The specimens should be stored in paper packets of good quality and uniform size (about 4 × 6 inches). Precise locality and meaningful habitat data, as well as the collector's name and date of collection, should be recorded in a neat and permanent fashion on an attached label. The packets can be filed in boxes or attached as many as eight to a sheet of herbarium paper.

Useful Literature

Warnstorf's account of North American peat mosses, published nearly a century past (266), is pleasantly informative. Blomquist's illustrated guide to the peat mosses of the southeastern United States (20) gives a straightforward explanation of the genus and its taxonomic structure. Excellent photographs of diagnostic features were published by Vitt and Andrus (258) and also by Lange (130). Regional treatments of European *Sphagna* can be used with profit in North America too. These include especially Nyholm's beautifully illustrated presentation of the Fennoscandian species (165), Hill's account of the British representatives of the genus (109), and Daniels and Eddy's *Handbook of European Sphagna* (53). Useful American treatments have been published by Andrus (8), Vitt and Andrus (258), Crum (44, 45, 46), and Crum and Anderson (47). The illustrations in Crum's revision of the North American peat mosses, of 1984 (45), some of them reproduced in this work, make comparisons of closely related species particularly easy.

The nomenclatural starting point for true mosses is quite reasonably set at Hedwig's *Species Muscorum* of 1801. *Sphagnum*, on the other hand, begins half a century earlier with Linnaeus' *Species Plantarum* of 1753. Linnaeus recognized only one species and two varieties of *Sphagnum*, and nearly a century later William Joseph

Hooker, otherwise admirable as a bryologist, said that there is probably only one species. However, during that century Ehrhart, Hedwig, Bridel, Schwaegrichen, and Müller recognized many additional species and contributed to a fund of knowledge that Schimper brought together and made understandable. His memoir on developmental morphology and taxonomy, published in French in 1857 (217) and, slightly altered, in German in 1858 (218), commands the greatest respect. His copious illustrations are unsurpassed in beauty.

The last half of the nineteenth century was a period of exploration and territorial aggrandizement. The many scientific expeditions associated with land-grabbing brought together a great many species for piecemeal description and eventual reevaluation. Röll, Warnstorf, and Braithwaite were much concerned with the European peat mosses (and those happen to be widespread elsewhere in the Northern Hemisphere as well). In 1911 Warnstorf produced in a single volume an account of all the *Sphagna* thus far known for the entire world (342 species) (267). Ambitiously titled the *Sphagnologia Universalis*, this treatment includes detailed keys and descriptions (in Latin and German). Warnstorf's work deserves high praise, yet his species concepts were unrealistically narrow. Andrews, in 1913 (5), presented an excellent revision of the North American species. In reducing many of the species recognized by Warnstorf and thereby making a wide-ranging *Sphagnum* flora understandable, he was guided by unrealistically broad concepts. Both Warnstorf and Andrews sorted species without regard to any meaning they might have in nature. Since their time, sphagnologists have become more aware of species as meaningful biological entities. They tend to agree on what to recognize but not always how. Some authors use only specific designations, regardless of how "good" the species may be, while others assign some taxa to a lower-than-specific level.

Andrews produced two long series of notes over the four decades following his revision, one on the North American species (6), the other on Warnstorf's *Sphagnum* herbarium (7). Useful as they are, they suffer from his continued parsimony in species definitions. A recent contribution reflective of current points of view is Isoviita's treatment of the nomenclature and synonymy of the European taxa (115) (equally useful in North America). The *Index Muscorum* presents a listed account of all the species known from the world (about

185 species), up to the present or nearly so, but the synonymies are taken from literature sources of unequal value and cannot always be taken seriously.

Ruhland's account of morphology (213) and Paul's key to species (on a world basis) (181), both appearing in the second edition, volume 10, of Engler and Prantl's *Die natürlichen Pflanzenfamilien*, are very good indeed.

A Taxonomic Overview of North American Peat Mosses, Especially of the Great Lakes Region

The following key to sections is presented in the interest of grouping species by relationships and thereby simplifying the task of identification.

1. Cortical cells of stems and branches fibrillose; branch leaves broad, broadly pointed, and deeply hooded-concave, rough at back of the apex and denticulate along a marginal resorption furrow.   I. *Sphagnum*, p. 222
1. Cortical cells not fibrillose; branch leaves generally narrower and usually tapered to a narrowly truncate apex, neither hooded nor roughened at the apex, usually entire, only rarely with marginal resorption.
    2. Cortical cells of branches uniformly porose; stem leaves very small; branch leaves denticulate-margined by a resorption furrow. II. *Rigida*, p. 225
    2. Cortical cells dimorphous, some large, apically porose, and retort-shaped; branch leaves nearly always entire (except across the apex), bordered by linear cells.
        3. Plants aquatic; branch leaves with hyaline cells very long and scarcely broader than the green cells, without fibrils. Not known from the Great Lakes region.   V. *Isocladus*, p. 227
        3. Plants mostly not aquatic; branch leaves with hyaline cells shorter, distinctly broader than green cells, very rarely lacking fibrils.
            4. Stems without branches or with 1(-3) branches per fascicle; stem and branch leaves similar in structure; hyaline cells with thick, septumlike fibrils, without pores on the outer surface (or rarely with 1–4 rounded gaps), with numerous short, variously oriented membrane pleats. Not known from the Great Lakes region.
                         VII. *Hemitheca*, p. 233
            4. Stems with 3 or more branches per fascicle; stem and branch leaves differentiated; hyaline cells of branch leaves with fibrils scarcely septumlike, on the outer surface with pores numerous, gaps none, and membrane pleats none or few and longitudinal.
                5. Branch leaves commonly secund; pores of hyaline cells of branch leaves and often stem leaves as well with numerous commissural pores generally crowded in beadlike rows.
                             VIII. *Subsecunda*, p. 233
                5. Branch leaves not or rarely somewhat secund; pores of branch leaves fewer, not crowded in beadlike rows, the stem leaves normally without pores.

6. Branches in fascicles of 6–12; capitulum conspicuously large and dense; stem leaves very small; branch leaves widely recurved when dry.   IX. *Polyclada*, p. 235
6. Branches in fascicles of 5 or fewer; capitulum not conspicuously large; stem leaves usually not noticeably small; branch leaves not widely recurved when dry (though sometimes variously spreading or abruptly squarrose at the tips).
   7. Plants green, brown, or red; green cells of branch leaves exposed exclusively or more broadly on the inner surface.   X. *Acutifolia*, p. 236
   7. Plants variously colored but not red; green cells exposed equally or more broadly on the outer surface.
      8. Branch leaves broadly ovate and broadly truncate, with hyaline cells having pores in 3's at adjoining corners and green cells equally exposed on both surfaces. Not known from the Great Lakes region.   III. *Insulosa*, p. 226
      8. Branch leaves longer and more tapered, with pores not noticeably in 3's and green cells exposed exclusively or more broadly on the outer surface.
         9. Plants sometimes submerged, hyaline cells of stem leaves extensively resorbed on the inner surface (or less commonly on both surfaces across the apex or down the middle); branch leaves often wavy-margined when dry.   VI. *Cuspidata*, p. 228
         9. Plants not submerged; hyaline cells of stem leaves mostly resorbed on the outer surface below and on both surfaces at the apex; branch leaves not wavy-margined but variously spreading at the tips, at least when dry.   IV. *Squarrosa*, p. 226

Section I. *Sphagnum*

The branches are stout and plump with broad, hooded-concave leaves that are rough at the back of the apex owing to large, irregular membrane gaps. The margins are denticulate-bordered by a resorption furrow. The hyaline cells, below the leaf apex, have ringed, elliptic pores along the commissures on the outer surface; at adjacent angles the pores are commonly grouped in 3's (fig. 8-20), sometimes forming a common "vestibule." The stem leaves are broadly oblong-lingulate and finely fringed; their hyaline cells are extensively resorbed on the outer surface and sometimes also on the inner. The cortical cells of both stems and branches are porose and spirally fibrillose.

1. Plants normally red or red-tinged; green cells of branch leaves central and included; cells of the stem cortex with 1 pore (or occasionally 2–4 pores) at the outer surface.   1. *S. magellanicum*
1. Plants not red; green cells exposed on 1 or both surfaces; cells of the stem cortex normally with numerous pores.
   2. Plants pale green; green cells of branch leaves truncate-elliptic in section and narrowly exposed on both surfaces by thickened ends.   2. *S. centrale*
   2. Plants normally yellowish or brownish; green cells triangular to trapezoidal in section, exclusively or more broadly exposed on the inner surface.
      3. Stem leaves with hyaline cells undivided; green cells of branch leaves isosceles-triangular in section and hyaline cells smooth on the inner side walls.   3. *S. palustre*
      3. Stem leaves with hyaline cells commonly divided; green cells of branch leaves equilateral-triangular or trapezoidal in section and hyaline cells normally ornamented by papillae or coarse ridges on the inner side walls.
         4. Branch leaves with green cells trapezoidal in section; hyaline cells normally papillose on the inner side walls.   5. *S. papillosum*
         4. Branch leaves with green cells equilateral-triangular in section; hyaline cells normally coarsely ornamented by comb fibrils on the inner side walls.   4. *S. imbricatum*

1. *Sphagnum magellanicum* Brid. (figs. 8-56, 8-83*a–n*) normally grows in the open and develops a red color in response to sunlight. (Plants shaded by bog shrubs may be green, but the wood cylinder is red and the stem leaves pinkish.) The hyaline cells of the branch leaves are flat on both surfaces. The green cells are small, central, and entirely included. The cortical cells of the stem are usually uniporose.

This common and widespread species initiates hummocks in the poor fen zone of bog mats. It can be seen very near the water of an acid lake invading *Sphagnum papillosum* lawns. In the low shrub and black spruce zones it persists at lower and middle levels of the hummocks, usually in company with *S. capillifolium*.

2. *Sphagnum centrale* C. Jens. (figs. 8-57, 8-84*a–l*) is a whitish green carpet-former in depressions of *Thuja* swamps (or, southward, in the shrubby outer zone of peatland mats). The hyaline cells of the branch leaves are convex on both surfaces. The green cells in section appear to be central and included, but they are actually elongate and narrowly exposed on both surfaces by thickened ends. The cortical cells of the stem have 2–5 pores, rarely more.

3. *Sphagnum palustre* L. (fig. 8-85a–m) is common in southerly portions of the Great Lakes area growing in the shrub zone separating peatland mats from hardwood swamps. It is a species of relatively mineral-rich habitats. The plants may be green or, more commonly, yellowish. The hyaline cells of branch leaves are convex on the inner surface and bulging on the outer, and the green cells are isosceles-triangular in section with the base exposed on the inner surface. In the upper Great Lakes region, some confusion results from the fact that some forms of *S. papillosum* have papillae on the side walls of hyaline cells reduced or even lacking. However, *S. papillosum* has branch leaves with green cells trapezoidal in section and stem leaves with hyaline cells divided.

Sphagnum henryense Warnst., a Coastal Plain moss with a few upland stations in the East, has much more numerous pores on the outer surface of branch leaves, 5–14 in upper median regions and 16–26 in lower cells. The pores are large, ± rounded, and toward the base of the leaf often enclosed by looped or sigmoid fibrils. The side walls of hyaline cells bear an irregular network of ridges. Confusion with *S. palustre* is possible because the network of ridges can be poorly represented or even lacking and also because *S. palustre* sometimes has rather numerous rounded pores toward the sides of the leaf base (2–7 above, 10, rarely 13, below).

4. *Sphagnum imbricatum* Hornsch. has a scattered occurrence in the continental interior (including sites in Missouri, Illinois, and Manitoba), and it may occur somewhere in the Great Lakes region, perhaps in sedgy fens or alder thickets (places where the mineral content is relatively high). It tends to be yellow-brown, with much the same appearance as *S. palustre* and *S. papillosum*. Its branch leaves have green cells equilaterally triangular and broadly exposed on the inner surface. The hyaline cells have comblike fibrils projecting from their side walls. The comb fibrils are usually quite conspicuous, but sometimes they can be found only at the leaf base and rarely not at all. The hyaline cells of stem leaves are divided.

5. *Sphagnum papillosum* Lindb. (figs. 8-58, 8-86a–n) grows in golden-brown lawns under poor fen conditions, that is, in relatively wet and acid habitats under the influence of groundwater. At the margins of acid lakes it immediately succeeds *S. cuspidatum*, but in

the case of mineral-rich waters where a sedge mat develops, *S. papillosum* can be expected farther back in the poor fen zone where *S. cuspidatum* and sometimes also *S. majus* occupy shallow pools. It is, here again, successive to aquatic species, and it is followed by the red hummock-formers, *S. magellanicum* and *S. capillifolium*.

As seen in section, the green cells of the branch leaves are narrowly trapezoidal, with somewhat broader exposure on the inner surface. The hyaline cells are densely papillose at the sides. The papillae are normally conspicuous, but they can be poorly developed, sometimes present only at the leaf base and sometimes completely absent. Specimens lacking papillae can be separated from *S. palustre* by the division of stem leaf hyaline cells. In addition, the cortical cells of the stem are few (rarely more than 1–3 per cell in *S. papillosum*, 2–7, rarely 13, in *S. palustre*).

The loose-growing *Sphagnum papillosum* has been harvested by hand in northern Michigan and sold at high prices as Golden Moss, useful in floriculture where whole stems are desirable (as in orchid culture and floral baskets).

Section II. *Rigida*

The *Rigida* have uniformly porose cells of the branch cortex and branch leaves much larger than stem leaves and margined by a resorption furrow. The cortical cells of stems and branches lack fibrils, and those of the stem lack pores. The very small stem leaves are bluntly oblong or oblong-deltoid and scarcely bordered, and their hyaline cells lack divisions and fibrils. The branch leaves are broadly oblong-ovate and deeply concave but not hooded or roughened at the broadly truncate apex. The pores on the inner surface of branch leaves are grouped in 3's.

6. *Sphagnum compactum* DC (figs. 8-59, 8-87a–m) grows in low, pale green or brownish white cushions on sand at the outer margins of open bogs and jack pine–dominated poor fens. The stems are dark. The leaves, generally wide-spreading at the tips, may be concave-tapered to the extent that the broadly truncate apex is not readily seen. The upper hyaline cells of branch leaves have pseudopores and also open pores along their margins, and the green cells are small, central, and included.

Section III. *Insulosa*

This section, consisting of a single arctic species, *S. aongstroemii* C. Hartm. (fig. 8-88a–m), has no distinctive characters but rather a unique combination of features. The plants, with some resemblance to members of the section *Sphagnum*, but on a small scale, are loosely tufted and have an insipid yellow color. The stem cortex lacks fibrils, but some of its cells, or many of them, have a single pore at the upper end. The stem leaves are oblong-lingulate and somewhat fringed at a broadly rounded apex. The border of linear cells is abruptly broadened below. The hyaline cells, mostly divided, are largely resorbed on the outer surface and show large gaps toward the leaf apex and pleats toward the base. The cells of the branch cortex are in part retortlike. The branch leaves are broadly ovate and deeply concave with a rather broad, truncate apex and a border of linear cells. The green cells are truncately elliptic in section and narrowly exposed on both surfaces. The pores on the outer surface of the hyaline cells are ringed, elliptic, and grouped in 3's.

Section IV. *Squarrosa*

The *Squarrosa* grow in base-rich habitats. They have a large terminal bud and spreading or even squarrose branch leaf tips, at least when dry. The stem cortex lacks fibrils and pores. The stem leaves are oblong-elliptic and apically rounded. The hyaline cells, undivided and lacking fibrils, are extensively resorbed on the outer surface and toward the apex on the inner surface as well. For that reason the leaf apex is sievelike and irregularly erose. The cells of the branch cortex are in part retortlike. The branch leaves are bordered by linear cells and have green cells exposed exclusively or more broadly on the outer surface. The side walls of the hyaline cells are very finely, in fact, faintly, papillose. The pores on the outer surface of branch leaf hyaline cells are large, rounded, and thin-margined. Those on the inner surface are elliptic with distinct or even ringed margins.

1. Branch leaves abruptly narrowed and squarrose-spreading above an erect base; stems generally pale.   7. *S. squarrosum*
1. Branch leaves evenly tapered, erect or ± spreading at the tips (especially when dry); stems dark brown.   8. *S. teres*

7. *Sphagnum squarrosum* Crome (figs. 8-60, 8-89*a–m*) can be recognized at a glance because of its fresh, green color and conspicuously squarrose-spreading branch leaf tips. The branch leaves are abruptly narrowed and wide-spreading beyond a broad, oblong base. The stems are generally pale green or more or less red-brown in plants exposed to the sun. The plants grow loosely in depressions of *Thuja* swamps.

8. *Sphagnum teres* (Schimp.) Ångstr. (figs. 8-61, 8-90*a–m*) is more slender than *S. squarrosum*. It grows in small cushions at the margins of mineral-rich lakes in pioneering sedge mats and also in alder thickets. The plants are yellow to brown, with dark brown stems. The branch leaves are gradually narrowed from an elliptic base. The acumen is usually erect when moist but more or less spreading when dry. The pores on the outer surface of branch leaf hyaline cells are large, thin-margined, and often irregular—more nearly gaps than pores. (In *S. squarrosum* the pores, though large and thin-margined, are regular in outline.) The large terminal bud and soppy habitat aid in field identification, as does the rounded, sievelike apex of stem leaves (easily seen in profile when the capitulum is plucked from the stem).

Section V. *Isocladus*

Distinctive in the extreme, the section *Isocladus* consists of a single species, *S. macrophyllum* Brid., of the Atlantic coast. The plants are stout aquatics. They may be dark green, red-brown, or dark brown to brownish, with a silvery sheen when dry. The stem cortex has neither fibrils nor pores. The branch cortex has no retort cells, although an occasional cell may be apically porose. The branch leaves are long, crowded, and bristly-spreading on short, uniformly spreading-deflexed branches. Bordered by linear cells, they have long, narrow hyaline cells, not much differentiated from green cells and lacking fibrils.

The var. *macrophyllum* (figs. 8-62, 8-91*a–h*) has up to 20 large, round pores in a single row on the outer surface of each hyaline cell. The var. *floridanum* Aust. has as many as 60 small, rounded or slitlike pores in 2 irregular rows. The var. *floridanum* (fig. 8-92*e*) is dark green, narrow-leaved, and bristly in appearance, whereas the var. *macrophyllum* is dark brown or red-brown with broader, less spreading leaves. The var. *burinense* Maass (fig. 8-92*f*), of Newfoundland, has no pores at all.

Section VI. *Cuspidata*

The *Cuspidata* grow mainly in wet depressions, in some cases submerged. They tend to be yellowish to brown. The cortical cells of the stem are sometimes poorly differentiated; they lack pores and fibrils. The stem leaves are often broadly bordered at base, and their hyaline cells are extensively resorbed on the inner surface, sometimes also on the outer surface across the apex or in a V-shaped area down the middle. The branch cortex has some of its cells retortlike. The branch leaves, bordered by linear cells and often wavy-margined when dry, have green cells exclusively or more broadly exposed on the outer surface.

1. Stem leaves extensively resorbed in upper median regions and often torn down the middle; terminal bud large and conic; tips of branch leaves consisting of uniformly green cells.   13. *S. riparium*
1. Stem leaves sometimes ± fringed but not extensively resorbed or deeply torn in the median region; terminal bud not or only moderately large; cells at tips of branch leaves dimorphous.
   2. Hyaline cells of branch leaves not fibrillose.   10. *S. splendens*
   2. Hyaline cells of branch leaves fibrillose.
      3. Plants not submerged.
         4. Hyaline cells on the outer surface near the apex of branch leaves with few pores but some connecting fibrils, on both surfaces toward the base with minute thin spots (visible on strong staining).   12. *S. obtusum*
         4. Hyaline cells of branch leaves with conspicuous window pores at upper ends, without connecting fibrils and thin spots.
            5. Plants green to yellowish or yellow-brown; young pendent branches (as seen between the rays of the capitulum) seeming to be paired; branch leaves not secund or noticeably in rows, oblong-lanceolate, with margins not reflexed when dry; green cells isosceles-triangular in section with the apex usually reaching the inner surface.   9. *S. recurvum*
            5. Plants generally brown; young pendent branches not paired; branch leaves often loosely secund, often in rows when moist, broadly ovate, with margins irregularly reflexed when dry; green cells of branch leaves equilateral-triangular with the apex about midway between the inner and outer surfaces.   11. *S. pulchrum*

3. Plants submerged (or stranded).
    6. Hyaline cells of branch leaves with pores on the outer surface few and small.
        7. Plants slender, flaccid, not plumose, green or yellow; terminal bud not noticeable; hyaline cells of branch leaves with pores on the inner surface 4, 5, or more. 14. *S. cuspidatum*
        7. Plants stout, ± rigid, with long, spreading branches, appearing spiny-plumose when submerged, dark green or brown; terminal bud rather prominent; hyaline cells of branch leaves with pores on the inner surface none or few. 15. *S. torreyanum*
    6. Hyaline cells of branch leaves with pores on the outer surface numerous.
        8. Capitulum relatively broad; hyaline cells of branch leaves with numerous large pores in 1–2 rows on the outer surface and few or none on the inner. 16. *S. majus*
        8. Capitulum small relative to the size of the plants; hyaline cells of branch leaves with an abundance of small pores on both surfaces. 17. *S. jensenii*

9. *Sphagnum recurvum* P.-Beauv. (figs. 8-63–66, 8-93*a–m*) is green in the shade but tinged with yellow-brown in the sun. It grows in hollows in low-shrub mats and sometimes mingled with *S. capillifolium* at lower levels of hummocks. It also occurs in loose lawns in woodland habitats marginal to the mat. The plants become whitish at the branch tips because of quick drying. Also on drying, the branch leaves become flattened out, reflexed at the tips, and wavy at the margins. The pores on the outer surface of branch leaf hyaline cells are small and inconspicuous except that the apical pore is rather larger and windowlike owing to a matching perforation on the opposite surface. The green cells are isosceles-triangular with the apex of the triangle usually reaching the inner surface of the leaf. The stems and branch bases are often flushed with pink.

Three troublesome segregates recognized primarily on the basis of stem leaf characters show some geographic differentiation but no correlation with habitat differences.

The var. *recurvum* is relatively stout, with stem leaves rather large, oblong-deltoid, and fringed across a broadly truncate apex. This variety is commonly called *S. flexuosum* Dozy & Molk. It has its best expression along the coast in the Southeast and in tropical America. It is uncommon inland.

The var. *brevifolium* (Braithw.) Warnst., also known as *S. fallax*

(Klinggr.) Klinggr., is rather large or medium-sized, with stem leaves deltoid and concave-acute or concave-apiculate. It is the most common expression in the Great Lakes region.

The var. *tenue* Klinggr. is slender, with small capitula, and its branch leaves are spreading-reflexed at the tips but not always wavy at the margins when dry. It often occupies relatively dry sites, crowded in hummocks in mixture with other species and perhaps for that reason limited in growth potential. Variability in size and branch leaf characteristics is such that the stem leaves need to be used for tongue-in-cheek identifications. They are small but broadly oblong-deltoid with a broad, flat, somewhat erose apex. This wide-ranging variety, often called *S. angustifolium* (Russ.) C. Jens., is common in the upper Great Lakes region and in a broad range farther north.

10. *Sphagnum splendens* Maass has been found but once, in Quebec, growing in mixture with *S. recurvum* var. *brevifolium*, from which it differs in its shiny dry aspect and the absence of fibrils in the hyaline cells of branch leaves (fig. 8-35).

11. *Sphagnum pulchrum* (Braithw.) Warnst. (figs. 8-67, 8-94a–l), in many ways similar to *S. recurvum*, is relatively robust and brownish, especially when dry. The spreading branches of the capitulum are stout and somewhat curved, with leaves often loosely curved-secund. The young pendent branches do not appear to be in pairs. The branches in and below the capitulum often have leaves in distinct rows when moist. On drying the leaves become shiny brown or yellow-brown and noticeably flattened out and broad, with margins variously reflexed and more or less wavy and tips abruptly concave-apiculate. The green cells of the branch leaves are equilaterally triangular in section, with the apex of the triangle falling short of the upper surface. *Sphagnum pulchrum* is uncommon in the region of the Great Lakes, where it prefers wet hollows of sedgy poor fens. Its association with *S. subsecundum* and *S. papillosum* suggests a pioneering role.

12. *Sphagnum obtusum* Warnst. (fig. 8-95a–n) is scattered across Canada and Alaska and has been found in northern Minnesota. It

can be expected in mineral-rich sedge mats. The plants are robust and have a greasy sheen when moist but otherwise resemble *S. recurvum*, having young pendent branches appearing to be paired and branch leaves undulate-margined when dry. The older spreading branches are relatively stout. The spreading branches of each young fascicle curve toward each other, giving the capitulum a curly aspect when viewed from above. However, the hyaline cells of the branch leaves lack apical window pores, at least in the upper portions of the leaf. On the outer surface near the leaf tips the hyaline cells have only a few small end and corner pores or sometimes fairly numerous small, nonringed, rounded pores or thin spots in addition to a fair number of ringed pseudopores and fibril connections. Toward the leaf base, and especially on the outer surface, is an abundance of tiny thin spots, mere pinpricks visible on strong staining, in 1 or 2 rows over the surface of the cells.

13. *Sphagnum riparium* Ångstr. (figs. 8-68–69, 8-96a–l) resembles *S. recurvum* and shows a similar habitat preference. It is found in the hollows of open, relatively acid portions of bog mats. It is not common. The plants are green and rather stout, with tumid branches and a large, conic terminal bud. The young pendent branches are not consistently paired. The stem leaves are resorbed in a narrow median region and often rent down the middle.

14. *Sphagnum cuspidatum* Hoffm. (figs. 8-72, 8-97a–l) grows submerged at the margins of acid bog lakes and also in shallow pools in *Sphagnum* lawns formed external to sedge mats at the margins of circumneutral lakes. As the pools (or drainage tracks) dry out during the summer, the green or yellow mats become emergent and take on the sodden appearance of wet cat fur. This drowned kitten phase often passes into a whitish border at the margin of a depression where the plants tend to dry out (because of a poor water-holding capacity), and the entire depression usually dries before the end of summer. When submerged the plants are weak and flaccid, but when removed from water they seem to take on substance and rigidity. The spreading branches slenderly taper to a point. When dry the branch leaves seem long and narrow, especially at branch tips, and somewhat wavy-margined. The hyaline cells of the branch leaves have only a few small, inconspicuous pores on the outer surface. On

the inner surface they are larger and more numerous, though thin-walled and often inconspicuous.

Confusion with *S. recurvum* should not be a problem. *Sphagnum recurvum* does not grow in wet *Sphagnum* lawns or submerged, and its young pendent branches seem paired.

15. *Sphagnum torreyanum* Sull. (fig. 8-73) is a large, stiff aquatic. The plants are strikingly plumose with bristly, burrlike capitula sunken just below the surface (or becoming emergent). The terminal bud is moderately prominent. The branch leaves are much larger and the pores on the inner surface of hyaline cells fewer than in *S. cuspidatum*. This species of the Atlantic coast is also known from a few inland localities in New York, Pennsylvania, and northern Michigan (in relatively acid water at the edge of a pioneering *Chamaedaphne* mat at Rexton Lake, Mackinac County).

16. *Sphagnum majus* (Russ.) C. Jens. (figs. 8-70–71, 8-98*a*–*m*) is relatively uncommon in the region of the Great Lakes. It grows submerged in habitats similar to those of *S. cuspidatum* and sometimes grows with that species in open poor fens and wet *Sphagnum* lawns. However, it favors somewhat more mineral-rich pools and drainage tracks. As compared with *S. cuspidatum* the plants are stouter and darker. The branches of the capitulum tend to be curved and give a curly appearance when viewed from above. The capitula tend to be uniformly brown, whereas in *S. cuspidatum* the green or yellow capitula may be mottled by dark branch bases. The branches are less slenderly tapered. The hyaline cells of branch leaves have numerous large, round pores on the outer surface, in a single median row or 2 irregular rows. Pores are very few or none on the inner surface.

17. *Sphagnum jensenii* H. Lindb. (fig. 8-99*a*–*m*), also known as *S. annulatum* var. *porosum* (Warnst.) Maass, is rare in the region of the Great Lakes, where it grows submerged at the margins of acid lakes. The plants are considerably longer and stouter than in *S. majus* and dark green or brownish, often with a surprisingly small capitulum but quite long branches in widely spaced fascicles. The terminal bud is relatively large. The branch leaves, somewhat wavy-margined

when dry, have hyaline cells with numerous small pores in 1 or 2 rows on both surfaces.

Section VII. *Hemitheca*

Consisting of a single species of the Atlantic coast and a few localities in the eastern mountains, *Sphagnum pylaesii* Brid. (fig. 8-100a–k), this section has many unique properties. The pseudopodia are very short, and the small capsules are immersed among scarcely differentiated perichaetial leaves. The capsule wall lacks pseudostomata. Plants growing in wet depressions or shallow pools are maroon to blackish and branched (with branches single rather than fascicled), whereas those of wet rock crevices are orange-yellow and essentially unbranched. (These expressions are so distinctive that one wonders if they would not retain their nature under uniform conditions of culture.) The stem cortex is 1–2-layered, and the branch cortex has only a slight differentiation of retort cells. The stem and branch leaves are similar in structure. The border is weak, and the hyaline cells have strong, doughnutlike fibrils (almost septumlike), short membrane pleats oriented in various directions in each subdivision of the cell, and pores none or very few (but sometimes, in seepage forms, with 1–5 irregularly rounded gaps in each cell).

Section VIII. *Subsecunda*

The plants grow in low, loose carpets or small cushions in wet, relatively base-rich habitats. They tend to be orange-yellow. The cells of the stem cortex (in 1–several layers) lack fibrils but sometimes have a single round pore, porelike thin spot, or horizontal crack at the upper end. The branch cortex has some cells retortlike. The branch leaves, commonly secund, are bordered by linear cells, and the hyaline cells on their outer surfaces generally have numerous elliptic pores in a beadlike arrangement along the commissures.

The examples of the Great Lakes region are keyed here as varietal expressions of *S. subsecundum*. Even though some case can be presented for specific rank for some of them, the variability over a broader range necessitates taxonomic caution until further studies are made.

1. Stem cortex in 1 layer.
    2. Stem leaves much shorter than branch leaves, with pores and sometimes fibrils in the upper one-fourth or less.   Var. *subsecundum*
    2. Stem leaves one-half or more the length of the branch leaves, fibrillose and porose in the upper one-third or nearly throughout.
        3. Hyaline cells of stem leaves with numerous pores on the outer surface, few to numerous on the inner.   Var. *rufescens*
        3. Hyaline cells with pores few or none on the outer surface, numerous on the inner.   Var. *inundatum*
1. Stem cortex in 2–3 layers (best observed unstained).
    4. Terminal bud large; branches not clearly differentiated into spreading and pendent types, not concealing the stem; stem leaves often as long as the branch leaves, sometimes longer, broadly elliptic.   Var. *platyphyllum*
    4. Terminal bud small; pendent branches concealing the stem; stem leaves small, deltoid to ± oblong.   Var. *contortum*

18. *Sphagnum subsecundum* Sturm var. *subsecundum* (figs. 8-74, 8-101*a–o*) grows in mineral-rich, open habitats, in sedge mats marginal to eutrophic lakes and also hollows of sedgy poor fens. The plants are usually yellow-green, yellow-brown, or orange, with a shiny-varnished appearance when dry. The branch leaves are curved-secund, especially on short, curved branches of the capitulum. The stems are usually quite dark, and the cortical cells of the stem are 1-layered. The stem leaves, considerably shorter than the branch leaves, are blunt and concave at the tip, with pores and fibrils (if any) only near the apex; the pores are normally restricted to the inner surface, but sometimes a few can be seen on the outer surface of a few apical cells as well.

The var. *subsecundum* gives little trouble in identification, but a number of satellite expressions, or varieties, occupying similar habitats, may be troublesome, perhaps because of variations related to seasonal fluctuations in the environment.

The var. *rufescens* (Nees & Hornsch.) Hüb. (fig. 8-75), also known as *S. lescurii* Sull., has leaves varying from broad and imbricate-erect (giving the branches a tumid appearance) to narrower and often curved-secund. The stem cortex is 1-layered. The oblong-lingulate stem leaves are more than half as long as the branch leaves and sometimes longer. Their hyaline cells, often divided, are fibrillose in the upper third to half or more and abundantly porose on the outer surface, often on the inner as well.

The var. *inundatum* (Russ.) C. Jens. is fairly robust and barely distinct from some forms of the var. *rufescens*, but its stem leaves are more clearly differentiated from the branch leaves. They are only about half as long as the branch leaves, with hyaline cells commonly divided and fibrillose in the upper third or somewhat more; the pores are numerous, though scattered (along the commissures) on the inner surface but few or none on the outer. The branch leaves may be erect, spreading, or secund.

The var. *platyphyllum* (Braithw.) Card., also known as *S. platyphyllum* (Braithw.) Warnst., has a stem cortex of 2–3 layers, a large terminal bud, and tumid, sausagelike branches that are scarcely differentiated into pendent and spreading types. The stem leaves are spreading, hooded-concave, and large, and they resemble the branch leaves in being fibrillose and porose throughout, with pores more numerous on the outer surface. The branch leaves are straight and imbricate rather than secund.

The var. *contortum* (Schultz) Hüb., also known as *S. contortum* Schultz, has stems with 2–3 layers of cortical cells. The stem leaves are shorter than branch leaves and have fibrils and pores in the upper one-fourth or more; the pores are few or none on the outer surface, but on the inner surface, toward the apex, they are few to numerous (and usually more numerous than on the outer surface). The branch leaves are usually secund; the pores are usually numerous on the outer surface but often arranged in interrupted rows.

Section IX. *Polyclada*

This section is immediately recognizable because of a shaggy, globose capitulum and top-heavy appearance. The dark and woody stems are wiry-flexuose when moist. The stem cortex has neither pores nor fibrils, and the stem leaves are small and lingulate. The branches are crowded in fascicles of 6–12, and the branch cortex includes retort cells. The branch leaves, bordered by linear cells, are neatly turned out and downward when dry, and their hyaline cells have, on their outer surfaces, strongly ringed, elliptic commissural pores. The green cells are more or less equally exposed on both surfaces. The section consists of a single species.

19. *Sphagnum wulfianum* Girg. (figs. 8-76, 8-102*a–l*) grows almost exclusively in *Thuja* swamps in loose, brownish green mounds

or ridges representing stumps and logs in an advanced state of decay. The species is recognizable at a distance of several paces because of its loose, shaggy, top-heavy look.

Section X. *Acutifolia*

The plants are compactly tufted and sometimes red or red-tinged. The stem cortex has no fibrils, but pores are sometimes present. The stem leaves are commonly abruptly broad-bordered below, and the hyaline cells, often divided, show considerable resorption on the inner surface. The branch cortex includes well-formed retort cells. The branch leaves are bordered by linear cells; the hyaline cells have elliptic, ringed commissural pores on the outer surface; and the green cells are triangular to trapezoidal with exposure exclusively or more broadly on the inner surface.

1. Cells of the stem cortex normally porose.
    2. Stem leaves fan-shaped, lacerate-fringed nearly all around.
    28. *S. fimbriatum*
    2. Stem leaves not fan-shaped, less conspicuously fringed across the apex or merely erose at the middle of the apex.
        3. Plants often tinged with red or red-mottled; stem leaves moderately erose at the middle of a rounded apex; hyaline cells of stem leaves near the apex short and undivided, with membrane pleats, the midbasal cells not conspicuously enlarged. 26. *S. russowii*
        3. Plants green; stem leaves coarsely erose-dentate across a broadly truncate apex; upper hyaline cells of stem leaves sometimes 1–2-divided, largely resorbed on both surfaces and lacking membrane pleats; midbasal hyaline cells of stem leaves conspicuously large in a triangular, sievelike area of resorption. 27. *S. girgensohnii*
1. Cells of the stem cortex not porose (or, in *S. quinquefarium*, with pores few and scattered).
    4. Hyaline cells of branch leaves with minute, strongly ringed pores on the outer surface. 21. *S. warnstorfii*
    4. Hyaline cells of branch leaves with large, elliptic, less strongly ringed pores on the outer surface.
        5. Stem leaves flat, oblong-lingulate, rounded at the apex and broadly bordered at the base.
            6. Plants brown; branches of the capitulum straight, with crowded, erect, narrow leaves; stem leaves rather longer than broad.
            22. *S. fuscum*
            6. Plants red; branches of the capitulum often upcurved, with leaves usually not particularly crowded, subsecund, broadly concave; stem leaves broadly lingulate. 20. *S. capillifolium* var. *tenellum*

5. Stem leaves oblong to oblong-triangular, broadly acute, concave-acute, or abruptly concave-apiculate, with the border often conspicuously broad at base.
   7. Stem leaves oblong-ovate to oblong-triangular, often about as broad as long, usually abruptly concave-apiculate, broadly bordered above and conspicuously broad-bordered at base; spreading branches normally in 3's; branch leaves normally 5-ranked and spreading at the tips when dry. 25. *S. quinquefarium*
   7. Stem leaves oblong to oblong-ovate, longer than broad, acute or concave-pointed to apiculate, narrowly bordered above, not conspicuously broad-bordered below; spreading branches in 2's; branch leaves not ranked when dry.
      8. Plants red or red-tinged, not shiny when dry; hyaline cells of stem leaves fibrillose on the outer surface and some few of them with large, irregularly rounded or oblong gaps, without pleats. 20. *S. capillifolium*
      8. Plants uniformly pale brown or yellow-brown and flecked with tinges of pink, violet-red, or purple, ± shiny when dry; hyaline cells normally without fibrils or with only remnants of them, without gaps on the outer surface, commonly with pleats.
         9. Plants yellow-brown, spotted with pink-, violet-, or purple-brown, shiny when dry; stem leaves broadly acute and abruptly concave-cuspidate, with the border only slightly broadened at base; branch leaves ± recurved when dry, oblong-lanceolate, with relatively long, narrow tips. 23. *S. subnitens*
         9. Plants uniformly pale brown, ± shiny when dry; stem leaves broadly acute, not concave-cuspidate, with the border moderately broadened at base; branch leaves appressed, oblong-ovate, and broadly short-acuminate. 24. *S. subfulvum*

20. *Sphagnum capillifolium* (Ehrh.) Hedw. (figs. 8-77, 8-103*a–n*), often called *S. nemoreum* Scop., is a red species of open peatlands. It often initiates hummocks in wet *Sphagnum* lawns and persists at the lower and intermediate levels of hummocks in low-shrub and spruce muskeg zones. The species is common and widespread. Unfortunately, except by habitat, it is not easy to recognize in the field and is often confused with *S. warnstorfii* and *S. russowii*. *Sphagnum capillifolium* has nicely globose-rounded capitula, at least in older, dryer hummock sites, but its best characters are those of the stem leaves: the narrow, oblong-ovate, and concave-pointed leaves are not conspicuously broad-bordered at base, and the hyaline cells are fibrillose and often divided, a scattered few of them having moderately large, round or oblong gaps on the outer surface.

Occasional inland specimens showing narrow stem leaves with commissural pores on the outer surface of hyaline cells have been referred to as var. *tenerum* (Sull.) Crum. It is likely, however, that such examples of isophylly are habitat-induced and not actually the same as *S. tenerum* Sull. The latter is a well-marked species of the Atlantic Coastal Plain where *S. capillifolium* is not found and where no gradation toward it can be demonstrated. The problem is knowing where to draw the line, geographical or otherwise, between *S. tenerum* and mimic-forms of *S. capillifolium*.

The var. *tenellum* (Schimp.) Crum, often called *S. rubellum* Wils., is also widely distributed. It is variable in every respect and has no habitat niche to justify its separation from the var. *capillifolium*. However, at its best it has flat-topped capitula with curved branches and relatively short, broad, moderately concave branch leaves that are not much crowded and often spreading or subsecund. The stem leaves are flat, broadly oblong-lingulate and rounded at the apex. The border is conspicuously broadened at base, and the hyaline cells are mostly divided and lack gaps and usually fibrils on the outer surface. (In its colonizing form, as in wet *Sphagnum* lawns, *S. capillifolium* does not have the dainty pom-pom look that it develops higher and drier in a hummock. This wet-lawn expression is often called the var. *tenellum* or, at the specific level, *S. rubellum*. However, the nature of the capitula, whether flat or convex, is not as diagnostic as the structure of stem leaves in distinguishing the var. *capillifolium* from the var. *tenellum*.)

Cyrus McQueen, who has been particularly concerned with taxonomic problems in the relationship of *S. capillifolium*, has written us (June 6, 1986) concerning *S. capillifolium* and *S. rubellum:* "In my common garden setup, these two taxa were virtually indistinguishable in gametophyte morphology after 12 months. However, they were easily distinguished on the basis of spore morphology. Spores of both these taxa remained clearly distinguishable from sporophytes that developed completely within the common garden." For that reason, he considered *S. rubellum* better regarded as *S. capillifolium* var. *tenellum*. (Mark Hill [108] had already decided on the basis of numerical analysis that the taxa could not be maintained separately.) McQueen did find that *S. subtile* (Russ.) Warnst. maintained specific distinction from *S. capillifolium*. Its claims to distinction are, however, subtle, to say the least!

21. *Sphagnum warnstorfii* Russ. (figs. 8-78, 8-104*a–m*) grows in dense, low mounds in rich fens and loose carpets in mineral-rich *Thuja* swamps. Dark green in the shade, the plants take on a brilliant red or purple-red in the open. The branch leaves when dry are 5-ranked and out-turned at the tips. The hyaline cells have on the outer surface, especially toward the leaf apex, minute, round, strongly ringed pores along the commissures and often contiguous with them. Farther down on the leaf the pores are small but not minute and elliptic. The hyaline cells of stem leaves are nearly all divided and subdivided and have membrane pleats but normally no fibrils.

22. *Sphagnum fuscum* (Schimp.) Klinggr. (figs. 8-79, 8-105*a–l*) grows compactly in rusty to dark brown or sometimes green-brown variegated tufts, generally at the dry tops of older hummocks in acid peatlands, but in rich fens it forms large mounds, with the red *S. warnstorfii* in lower, wetter mounds. The habitat, color, compact growth, and slender, thready branches giving a wefted appearance to the lower portions aid in recognition. More tangible characters are provided by nearly flat and rather narrow stem leaves rounded at the apex, with the border strongly differentiated at base and hyaline cells 1–2-divided and normally lacking fibrils and pleats.

23. *Sphagnum subnitens* Warnst. (figs. 8-106*a–l*, 8-107*d–f*), rare inland, has been found in northern Michigan and also, reportedly, in Wisconsin. In oceanic areas it grows in bogs, but inland it prefers moderately minerotrophic habitats. The plants are soft and yellowish to brownish with pink or violet variegation and, when dry, have a blue-gray sheen. The stem leaves are pinched-cuspidate, and the hyaline cells are generally divided once, twice, or more. Fibrils are rarely present as mere traces, but membrane pleats are often seen. The branch leaves are narrow and slenderly involute-pointed.

24. *Sphagnum subfulvum* Sjörs (fig. 8-107*a–c*) is soft and tawny with only a slight sheen when dry. The capitulum is relatively broad, and the long branches subtending it have, when dry, a subtle shading of pink. The stem leaves are broadly oblong and narrowed to a somewhat concave, right-angled apex. The hyaline cells are mostly divided once or twice and commonly show a good development of membrane pleats and often traces of fibrils. The branch

leaves are relatively broad and broad-tapered. (Occasional loose-growing, orange- or yellow-brown forms of *S. fuscum* can be distinguished by a smaller size and narrower capitula with no pink tinges below them, as well as stem leaves more rounded at the apex and more strongly bordered at the base, with hyaline cells lacking any trace of fibrils or pleats.) *Sphagnum subfulvum* has been found in only a few sites in the Great Lakes region in depressions in sedgy poor fen habitats. It is contrasted in figure 8-107 with *S. subnitens* (*d–f*) and *S. flavicomans* (*g–l*), the latter coastal in distribution.

25. *Sphagnum quinquefarium* (Braithw.) Warnst. (fig. 8-108*a–m*), typically found in coniferous swamps (or, in montane areas, on steep slopes or cliffs), is known from a few inland areas, in Mackinac County, Michigan, and Cook County, Minnesota. The plants may be pale green, pink-tinged, or red, depending on degree of exposure. The branches are typically in fascicles of 3 spreading and 1 pendent. The branch leaves, when dry, are arranged in rows and have spreading tips. The stem cortex has a scattered few porose cells. The stem leaves are broadly oblong-triangular and abruptly concave-pointed; the border is broad above and conspicuously broadened at base.

26. *Sphagnum russowii* Warnst. (figs. 8-80, 8-109*a–l*) is similar to *S. capillifolium* and likewise difficult to recognize in the field. It forms cushions on older hummocks in low-shrub and black spruce zones. It also grows in *Thuja* swamps and on wooded slopes and moist or seepy cliffs. The plants are often reddish-tinged or conspicuously speckled by red antheridial catkins. The terminal bud is rather prominent, and the relatively coarse, blunt branches of the capitulum take on some familiarity to the practiced eye, but the best characters are microscopic: The cortical cells of the stem are generally porose, although pores are occasionally few and difficult to demonstrate. The stem leaves are somewhat erose at the middle of a rounded apex. They are bordered nearly all around and broadly so at the base. The stem leaves have apical cells short and undivided, with membrane pleats (but only rarely fibrils).

Well-developed swamp plants may be loosely tufted with star-shaped capitula and long, decurved branches as in *S. girgensohnii* but on a smaller scale.

27. *Sphagnum girgensohnii* Russ. (figs. 8-81, 8-110*a–l*) grows in large, loose carpets or mounds in *Thuja* swamps. The capitula are strikingly radiate, and the spreading branches are long and gracefully decurved in 5 rows. The terminal bud is fairly large. The squarish cortical cells of the stem are porose, and the stem leaves show considerable resorption in a triangular midbasal area and also across a broadly truncate, coarsely erose apex.

28. *Sphagnum fimbriatum* Wils. & J.D. Hook. (figs. 8-82, 8-111*a–l*) is slender, pale-green, and commonly fruited. It grows in bogs in loose mounds supported by lower branches of *Chamaedaphne* and also in small cushions in swampy woodlands (including both hardwood and tamarack swamps in the southern Great Lakes region). The plants are grayish green when dry, and the large and conspicuous terminal bud seems grayish-cobwebby because of the fringing of its component leaves. The branches of the capitulum are slender, and those below it flexuose-stringy. The cortical cells of the stem cortex are porose. The stem leaves are fan-shaped and remarkably fimbriate nearly all around. The broad, basal border encloses a sievelike, triangular area of resorption.

Fig. 8-56. *Sphagnum magellanicum*

Fig. 8-57. *S. centrale*

Fig. 8-58. *S. papillosum*

Fig. 8-59. *S. compactum*

Fig. 8-60. *S. squarrosum*

Fig. 8-61. *S. teres*

Figs. 8-63–66. *S. recurvum* var. *brevifolium*

Fig. 8-62. *Sphagnum macrophyllum.* (Photo by L. E. Anderson.)

Fig. 8-63. Habit

Fig. 8-64 **Fig. 8-65** **Fig. 8-66**

Figs. 8-64–66. Views from the side to show colored (brown) antheridial catkins and young pendent branches in pairs and also the appearance of branch leaves when dry

Fig. 8-67. *Sphagnum pulchrum.* (Photo by W. R. Buck.)

Fig. 8-69

Figs. 8-68–69. *S. riparium.* (Photo by Jeffrey Holcombe.)

Fig. 8-68

Fig. 8-70. *S. majus* submerged

Fig. 8-71. *S. majus* emergent

Fig. 8-72. *Sphagnum cuspidatum*, showing tapered branches

Fig. 8-74. *S. subsecundum*

Fig. 8-73. *S. torreyanum*

Fig. 8-75. *S. subsecundum* var. *rufescens*. In both *S. subsecundum* and the var. *rufescens*, there is variation in the curling of branches and secundity of leaves. (Photo by David Lane.)

Fig. 8-76. *S. wulfianum*

Fig. 8-77. *S. capillifolium* characteristically has nicely rounded capitula.

Fig. 8-78. *Sphagnum warnstorfii* is particularly slender. Growing in cedar swamps, it is dark green and not crowded. In the open, in rich fens, it may be brilliantly red and crowded in cushions. (The larger plants are *S. centrale*.)

Fig. 8-79. *S. fuscum* grows densely crowded in cushions, often at the top of old, dry hummocks. It is uniformly brown or brown mottled with green.

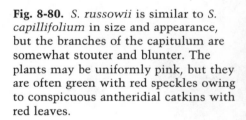

Fig. 8-80. *S. russowii* is similar to *S. capillifolium* in size and appearance, but the branches of the capitulum are somewhat stouter and blunter. The plants may be uniformly pink, but they are often green with red speckles owing to conspicuous antheridial catkins with red leaves.

Fig. 8-81. *S. girgensohnii* is relatively robust with branches 5-radiate and long-decurved (as viewed from the side).

Fig. 8-82. *S. fimbriatum* is very slender, with wispy, flexuous branches. It is almost always found in a fruiting condition. A large terminal bud, coupled with the slender size of the plants, aids in field recognition.

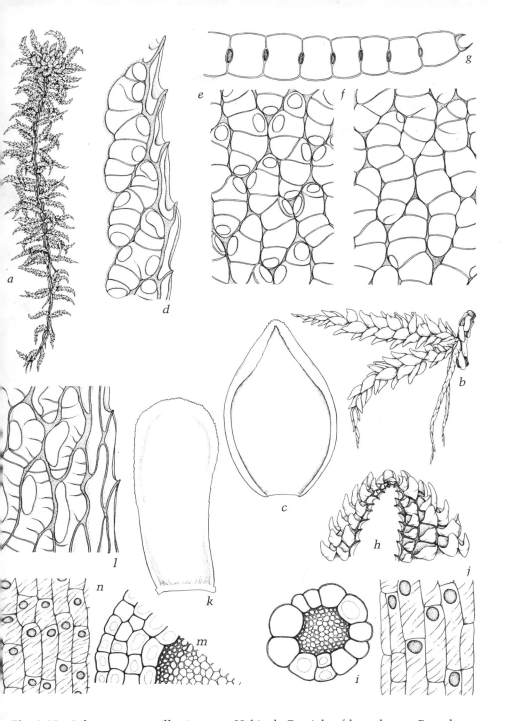

Fig. 8-83. *Sphagnum magellanicum.* *a.* Habit. *b.* Fascicle of branches. *c.* Branch leaf. *d.* Marginal resorption furrow. *e.* Upper cells of branch leaf, outer surface. *f.* Upper cells of branch leaf, inner surface. *g.* Cross-section of branch leaf. *h.* Apex of branch leaf showing roughness due to resorption. *i.* Cross-section of branch. *j.* Branch epidermis. *k.* Stem leaf. *l.* Stem leaf cells at margin. *m.* Cross-section of stem. *n.* Stem epidermis.

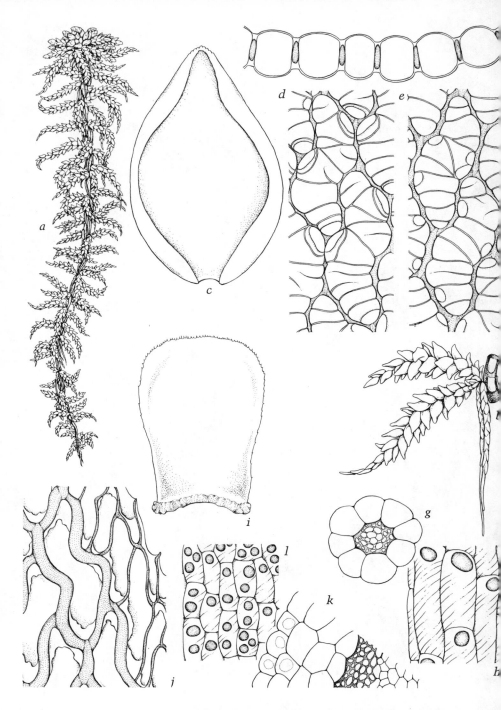

Fig. 8-84. *Sphagnum centrale. a.* Habit. *b.* Fascicle of branches. *c.* Branch leaf. *d.* Upper cells of branch leaf, outer surface. *e.* Upper cells of branch leaf, inner surface. *f.* Cross-section of branch leaf. *g.* Cross-section of branch. *h.* Branch epidermis. *i.* Stem leaf. *j.* Stem leaf cells at margin. *k.* Cross-section of stem. *l.* Stem epidermis.

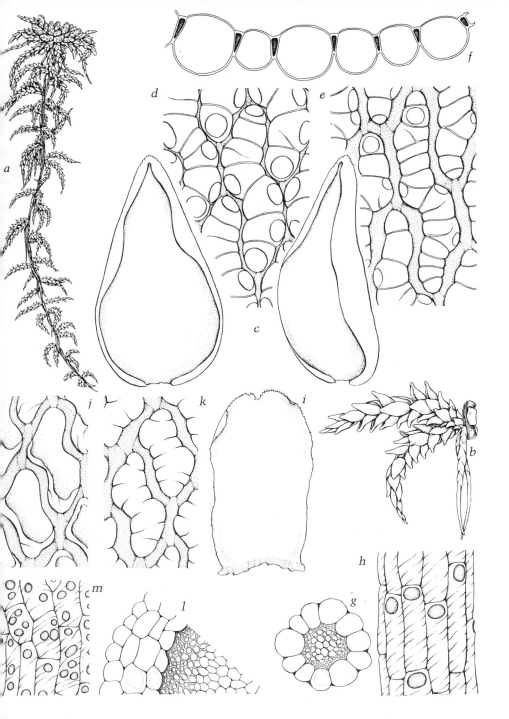

Fig. 8-85. *Sphagnum palustre. a.* Habit. *b.* Fascicle of branches. *c.* Branch leaves. *d.* Upper cells of branch leaf, outer surface. *e.* Upper cells of branch leaf, inner surface. *f.* Cross-section of branch leaf. *g.* Cross-section of branch. *h.* Branch epidermis. *i.* Stem leaf. *j.* Upper cells of stem leaf, outer surface. *k.* Upper cells of stem leaf, inner surface. *l.* Cross-section of stem. *m.* Stem epidermis.

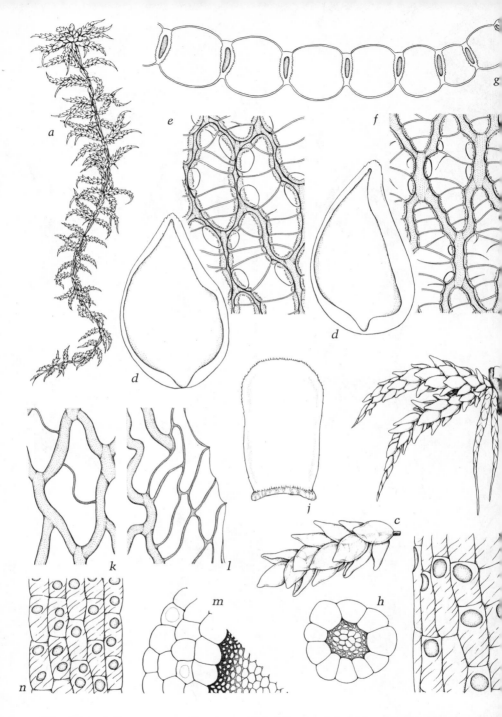

Fig. 8-86. *Sphagnum papillosum.* *a.* Habit. *b.* Fascicle of branches. *c.* Enlarged branch. *d.* Branch leaves. *e.* Upper cells of branch leaf, outer surface. *f.* Upper cells of branch leaf, inner surface. *g.* Cross-section of branch leaf. *h.* Cross-section of branch. *i.* Branch epidermis. *j.* Stem leaf. *k.* Upper cells of stem leaf. *l.* Upper marginal cells of stem leaf. *m.* Cross-section of stem. *n.* Stem epidermis.

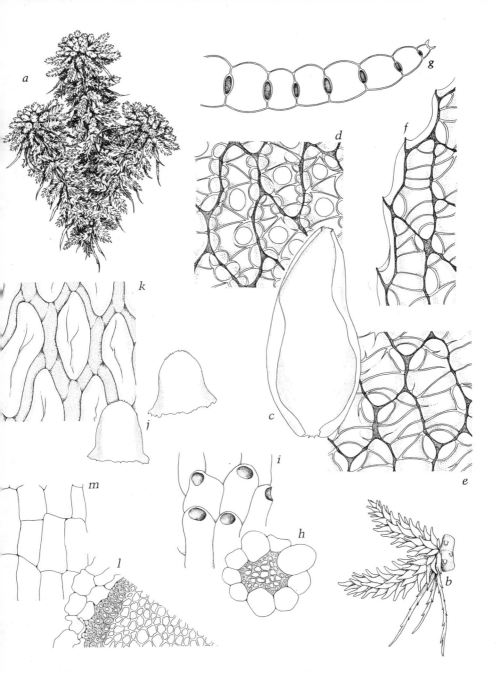

Fig. 8-87. *Sphagnum compactum.* *a.* Habit. *b.* Fascicle of branches. *c.* Branch leaf. *d.* Upper cells of branch leaf, outer surface. *e.* Upper cells of branch leaf, inner surface. *f.* Marginal resorption furrow of branch leaf. *g.* Cross-section of branch leaf. *h.* Cross-section of branch. *i.* Branch epidermis. *j.* Stem leaves. *k.* Stem leaf cells. *l.* Cross-section of stem. *m.* Stem epidermis.

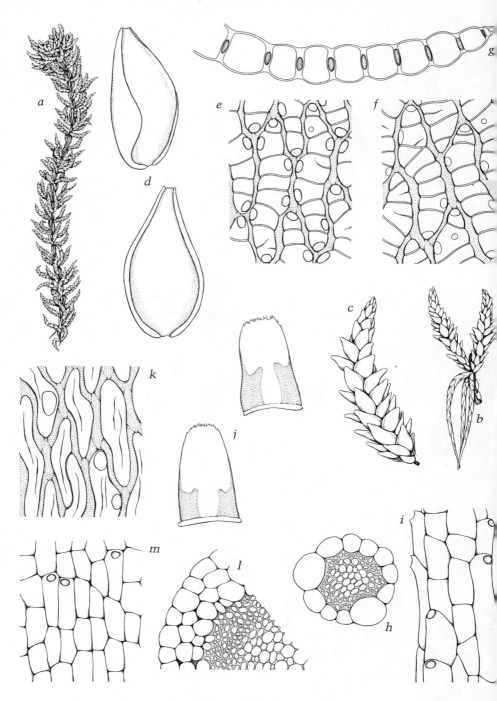

Fig. 8-88. *Sphagnum aongstroemii*. *a.* Habit. *b.* Fascicle of branches. *c.* Branch, enlarged. *d.* Branch leaves. *e.* Upper cells of branch leaf, outer surface. *f.* Upper cells of branch leaf, inner surface. *g.* Cross-section of branch leaf. *h.* Cross-section of branch. *i.* Branch epidermis. *j.* Stem leaves. *k.* Stem leaf cells. *l.* Cross-section of stem. *m.* Stem epidermis.

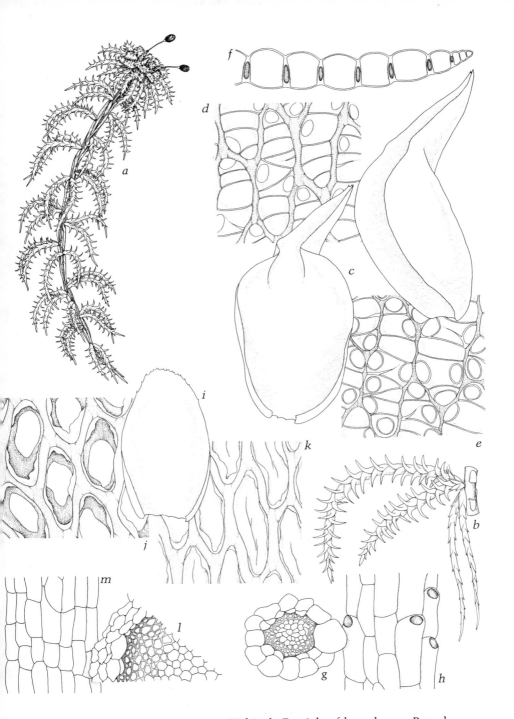

Fig. 8-89. *Sphagnum squarrosum.* a. Habit. b. Fascicle of branches. c. Branch leaves. d. Upper cells of branch leaf, outer surface. e. Upper cells of branch leaf, inner surface. f. Cross-section of branch leaf. g. Cross-section of branch. h. Branch epidermis. i. Stem leaf. j. Upper cells of stem leaf showing resorption on both surfaces. k. Upper median cells of stem leaf. l. Cross-section of stem. m. Stem epidermis.

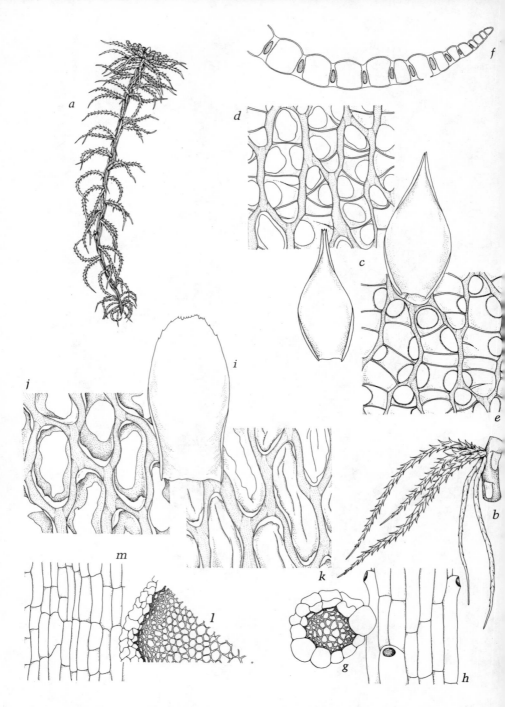

Fig. 8-90. *Sphagnum teres. a.* Habit. *b.* Fascicle of branches. *c.* Branch leaves. *d.* Upper cells of branch leaf, outer surface. *e.* Upper cells of branch leaf, inner surface. *f.* Cross-section of branch leaf. *g.* Cross-section of branch. *h.* Branch epidermis. *i.* Stem leaf. *j.* Upper cells of stem leaf showing resorption on both surfaces. *k.* Upper median cells of stem leaf, inner surface. *l.* Cross-section of stem. *m.* Stem epidermis.

Fig. 8-91. *Sphagnum macrophyllum.* *a.* Habit. *b.* Fascicle of branches. *c.* Stem leaves. *d.* Upper cells of stem leaf, inner surface. *e.* Cross-section of branch. *f.* Branch epidermis. *g.* Cross-section of stem. *h.* Stem epidermis.

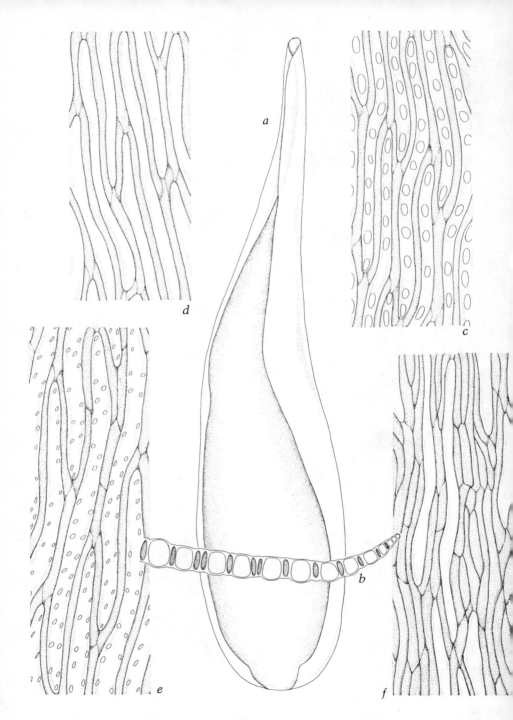

Fig. 8-92. *Sphagnum macrophyllum* (continued). *a.* Branch leaf. *b.* Cross-section of branch leaf. *c.* Branch leaf cells, outer surface. *d.* Branch leaf cells, inner surface. *e.* Var. *floridanum*, branch leaf cells, outer surface. *f.* Var. *burinense*, branch leaf cells, outer surface.

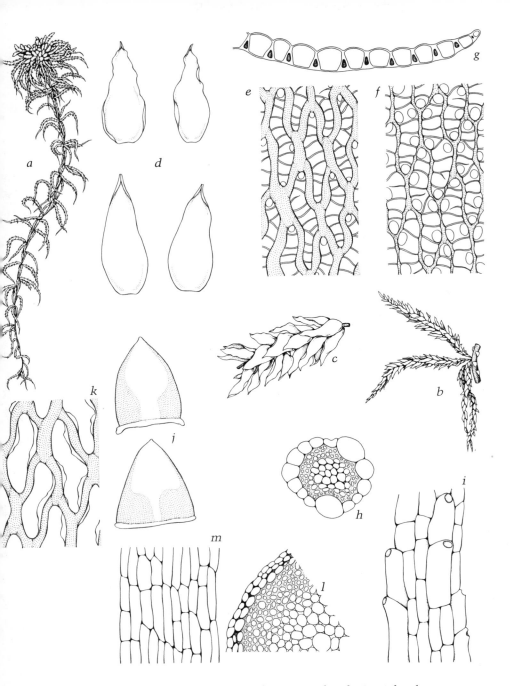

Fig. 8-93. *Sphagnum recurvum* var. *brevifolium*. *a*. Habit. *b*. Fascicle of branches. *c*. Branch, dry. *d*. Branch leaves, moist and dry. *e*. Upper cells of branch leaf, outer surface. *f*. Upper cells of branch leaf, inner surface. *g*. Cross-section of branch leaf. *h*. Cross-section of branch. *i*. Branch epidermis. *j*. Stem leaves. *k*. Upper cells of stem leaf. *l*. Cross-section of stem. *m*. Stem epidermis.

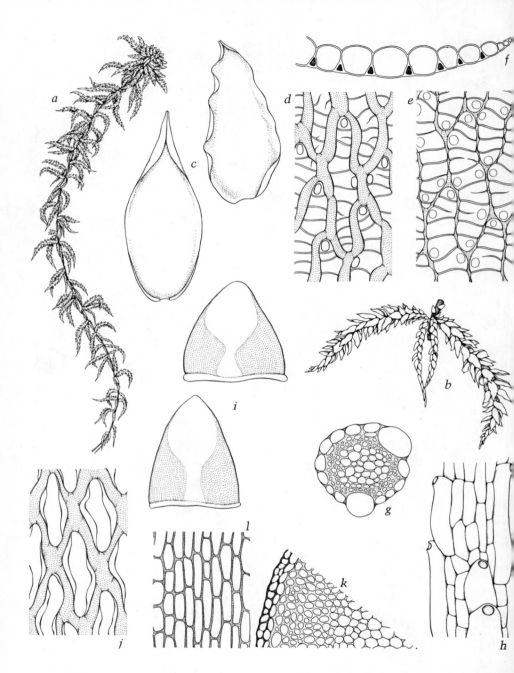

Fig. 8-94. *Sphagnum pulchrum. a.* Habitat. *b.* Fascicle of branches. *c.* Branch leaves, wet and dry. *d.* Upper cells of branch leaf, outer surface. *e.* Upper cells of branch leaf, inner surface. *f.* Cross-section of branch leaf. *g.* Cross-section of branch. *h.* Branch epidermis. *i.* Stem leaves. *j.* Upper cells of stem leaf, inner surface. *k.* Cross-section of stem. *l.* Stem epidermis.

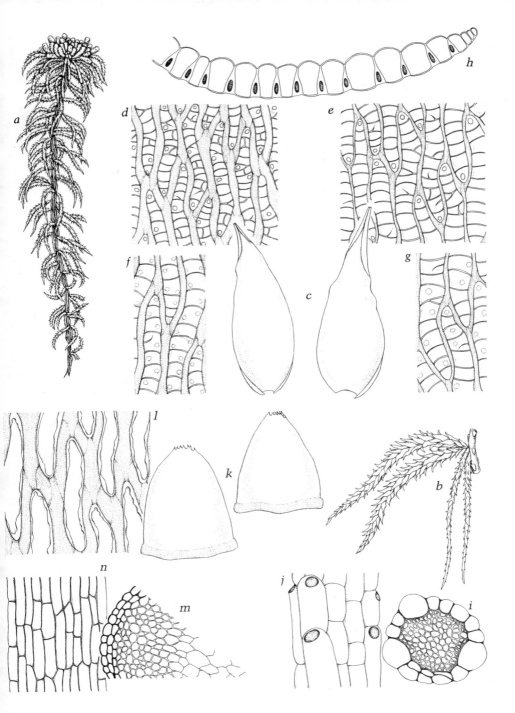

Fig. 8-95. *Sphagnum obtusum*. *a*. Habit. *b*. Fascicle of branches. *c*. Branch leaves. *d*. Upper cells of branch leaf, outer surface. *e*. Upper cells of branch leaf, inner surface. *f*. Lower cells of branch leaf, outer surface. *g*. Lower cells of branch leaf, inner surface. *h*. Cross-section of branch leaf. *i*. Cross-section of branch. *j*. Branch epidermis. *k*. Stem leaves. *l*. Upper cells of stem leaf, near apex, inner surface. *m*. Cross-section of stem. *n*. Stem epidermis.

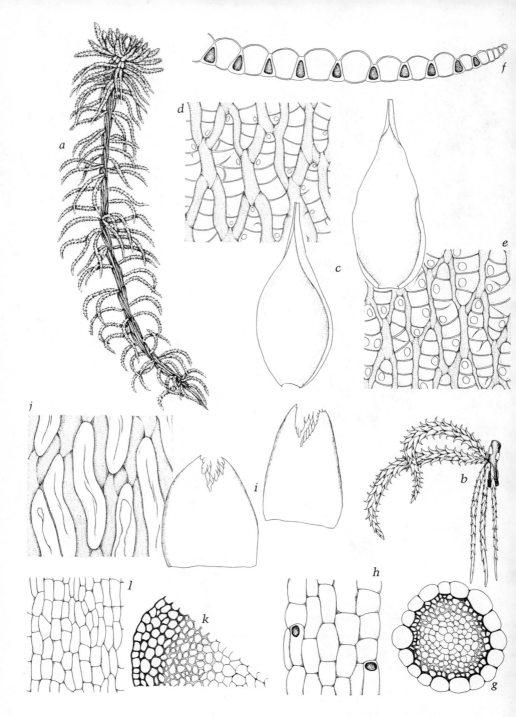

Fig. 8-96. *Sphagnum riparium.* *a.* Habit. *b.* Fascicle of branches. *c.* Branch leaves. *d.* Upper cells of branch leaf, outer surface. *e.* Upper cells of branch leaf, inner surface. *f.* Cross-section of branch leaf. *g.* Cross-section of branch. *h.* Branch epidermis. *i.* Stem leaves. *j.* Upper cells of stem leaf, outside of heavily resorbed median portion, outer surface. *k.* Cross-section of stem. *l.* Stem epidermis.

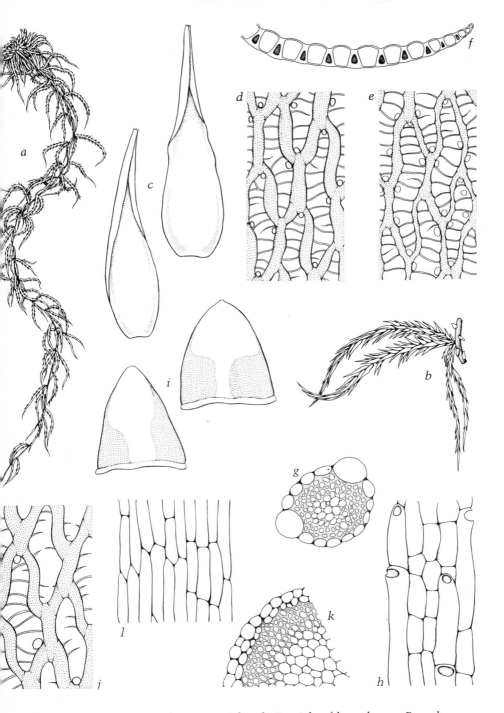

Fig. 8-97. *Sphagnum cuspidatum.* a. Habit. b. Fascicle of branches. c. Branch leaves. d. Upper cells of branch leaf, outer surface. e. Upper cells of branch leaf, inner surface. f. Cross-section of branch leaf. g. Cross-section of branch. h. Branch epidermis. i. Stem leaves. j. Upper cells of stem leaf, outer surface. k. Cross-section of stem. l. Stem epidermis.

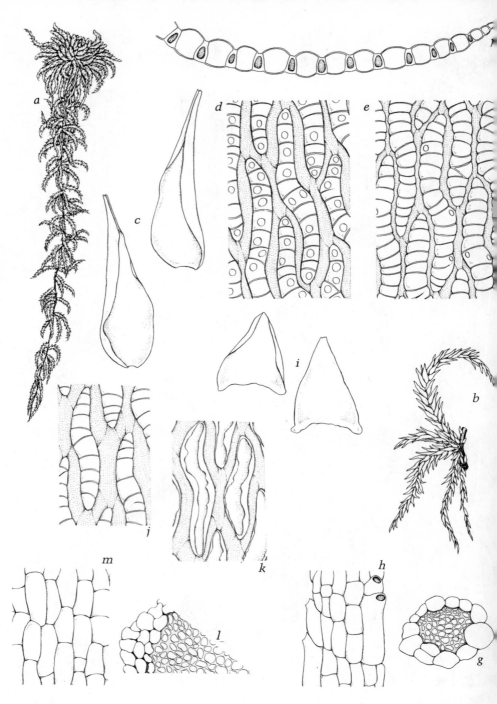

Fig. 8-98. *Sphagnum majus. a.* Habit. *b.* Fascicle of branches. *c.* Branch leaves. *d.* Upper cells of branch leaf, outer surface. *e.* Upper cells of branch leaf, inner surface. *f.* Cross-section of branch leaf. *g.* Cross-section of branch. *h.* Branch epidermis. *i.* Stem leaves. *j.* Upper cells of stem leaf, outer surface. *k.* Upper cells of branch leaf, inner surface. *l.* Cross-section of stem. *m.* Stem epidermis.

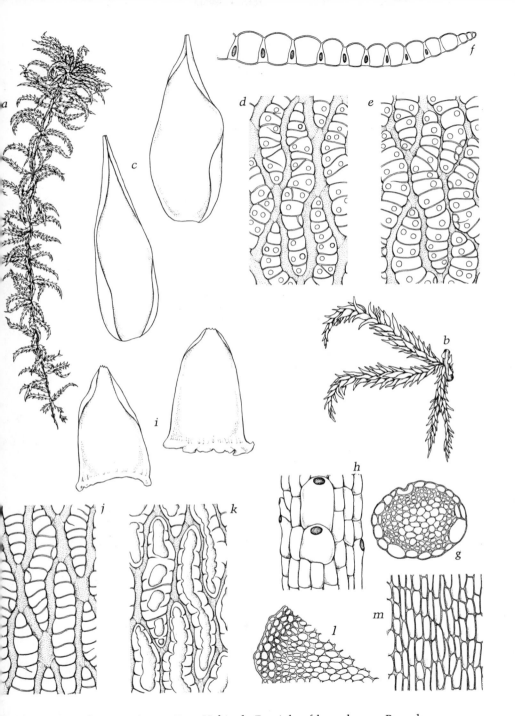

Fig. 8-99. *Sphagnum jensenii.* *a.* Habit. *b.* Fascicle of branches. *c.* Branch leaves. *d.* Upper cells of branch leaf, outer surface. *e.* Upper cells of branch leaf, inner surface. *f.* Cross-section of branch leaf. *g.* Cross-section of branch. *h.* Branch epidermis. *i.* Stem leaves. *j.* Upper cells of stem leaf, outer surface. *k.* Upper cells of stem leaf, inner surface. *l.* Cross-section of stem. *m.* Stem epidermis.

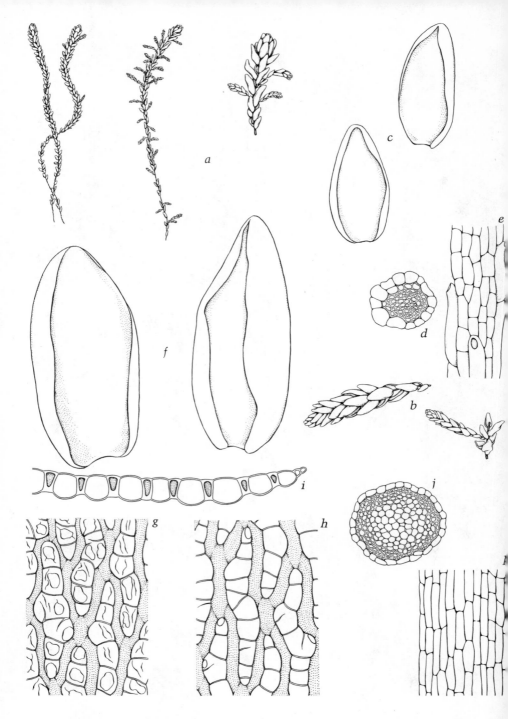

Fig. 8-100. *Sphagnum pylaesii.* *a.* Habits. *b.* Branches. *c.* Branch leaves. *d.* Cross-section of branch. *e.* Branch epidermis. *f.* Stem leaves. *g.* Stem leaf cells, outer surface. *h.* Stem leaf cells, inner surface. *i.* Cross-section of stem leaf. *j.* Cross-section of stem. *k.* Stem epidermis.

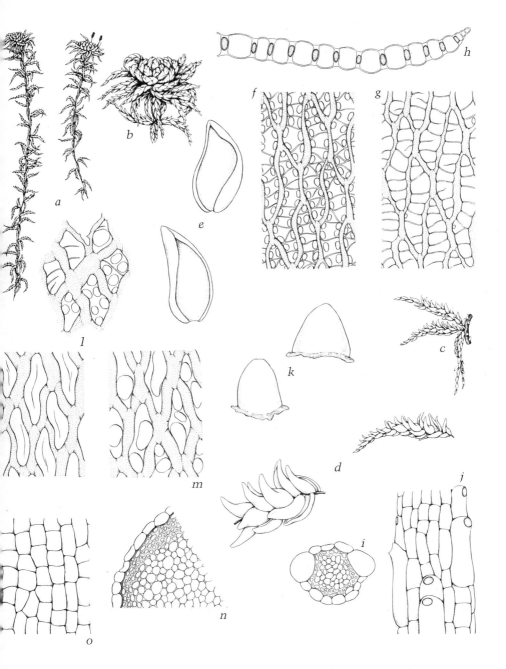

Fig. 8-101. *Sphagnum subsecundum.* *a.* Habits. *b.* Capitulum. *c.* Fascicle of branches. *d.* Branches, enlarged. *e.* Branch leaves. *f.* Upper cells of branch leaf, outer surface. *g.* Upper cells of branch leaf, inner surface. *h.* Cross-section of branch leaf. *i.* Cross-section of branch. *j.* Branch epidermis. *k.* Stem leaves. *l.* Upper cells of stem leaf, outer surface. *m.* Upper cells of stem leaf, inner surface. *n.* Cross-section of stem. *o.* Stem epidermis.

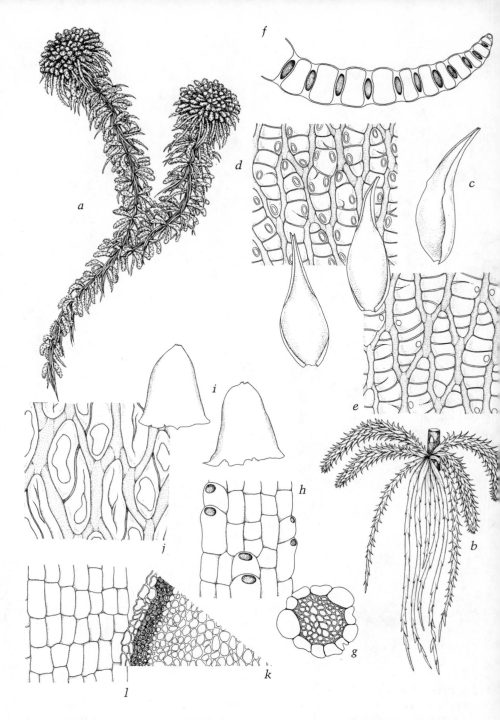

Fig. 8-102. *Sphagnum wulfianum*. a. Habit. b. Fascicle of branches. c. Branch leaves. d. Upper cells of branch leaf, outer surface. e. Upper cells of branch leaf, inner surface. f. Cross-section of branch leaf. g. Cross-section of branch. h. Branch epidermis. i. Stem leaves. j. Upper cells of stem leaf, outer surface. k. Cross-section of stem. l. Stem epidermis.

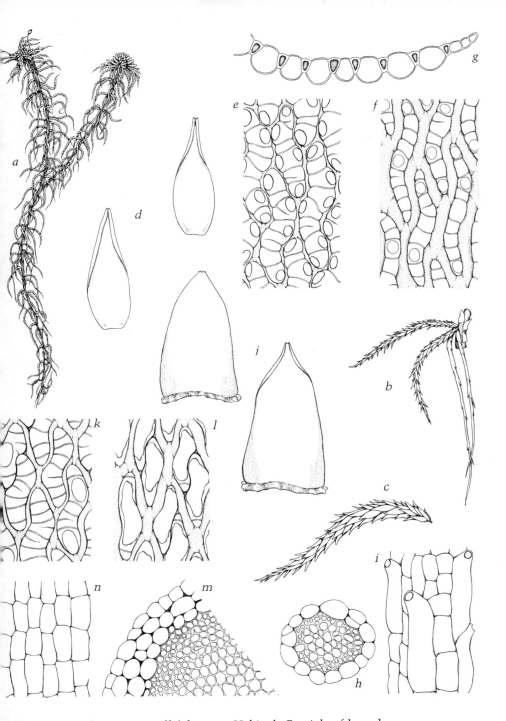

Fig. 8-103. *Sphagnum capillifolium.* *a.* Habit. *b.* Fascicle of branches. *c.* Branch, enlarged. *d.* Branch leaves. *e.* Upper cells of branch leaf, outer surface. *f.* Upper cells of branch leaf, inner surface. *g.* Cross-section of branch leaf. *h.* Cross-section of branch. *i.* Branch epidermis. *j.* Stem leaves. *k.* Upper cells of stem leaf, outer surface. *l.* Upper cells of stem leaf, inner surface. *m.* Cross-section of stem. *n.* Stem epidermis.

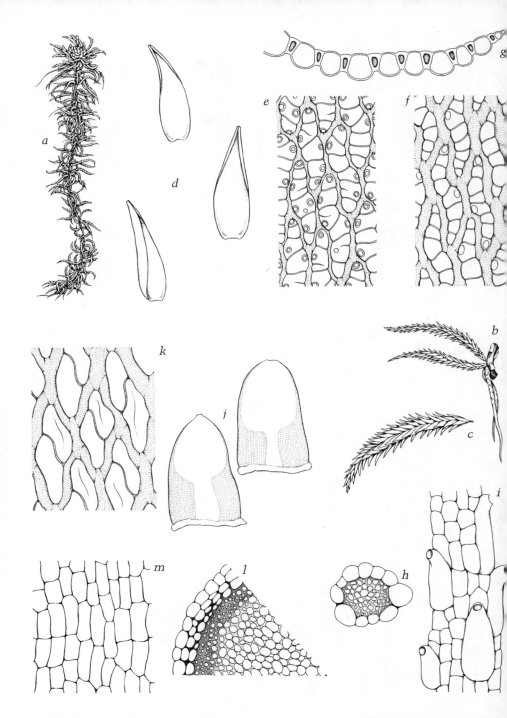

Fig. 8-104. *Sphagnum warnstorfii.* *a.* Habit. *b.* Fascicle of branches. *c.* Branch, enlarged. *d.* Branch leaves. *e.* Upper cells of branch leaf, outer surface. *f.* Upper cells of branch leaf, inner surface. *g.* Cross-section of branch leaf. *h.* Cross-section of branch. *i.* Branch epidermis. *j.* Stem leaves. *k.* Upper cells of stem leaf, outer surface. *l.* Cross-section of stem. *m.* Stem epidermis.

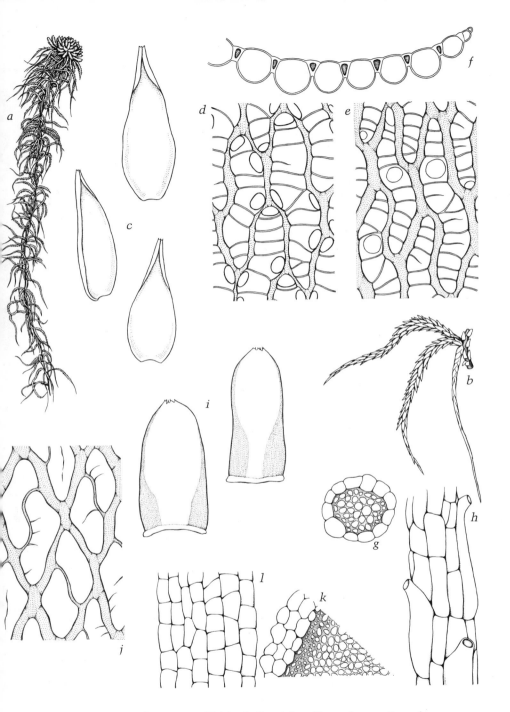

Fig. 8-105. *Sphagnum fuscum.* *a.* Habit. *b.* Fascicle of branches. *c.* Branch leaves. *d.* Upper cells of branch leaf, outer surface. *e.* Upper cells of branch leaf, inner surface. *f.* Cross-section of branch leaf. *g.* Cross-section of branch. *h.* Branch epidermis. *i.* Stem leaves. *j.* Upper cells of stem leaf. *k.* Cross-section of stem. *l.* Stem epidermis.

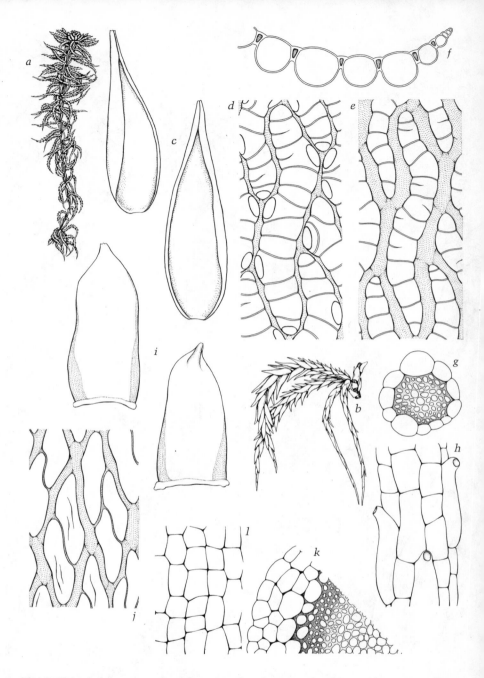

Fig. 8-106. *Sphagnum subnitens. a.* Habit. *b.* Fascicle of branches. *c.* Branch leaves. *d.* Upper cells of branch leaf, outer surface. *e.* Upper cells of branch leaf, inner surface. *f.* Cross-section of branch leaf. *g.* Cross-section of branch. *h.* Branch epidermis. *i.* Stem leaves. *j.* Upper cells of stem leaf, outer surface. *k.* Cross-section of stem. *l.* Stem epidermis.

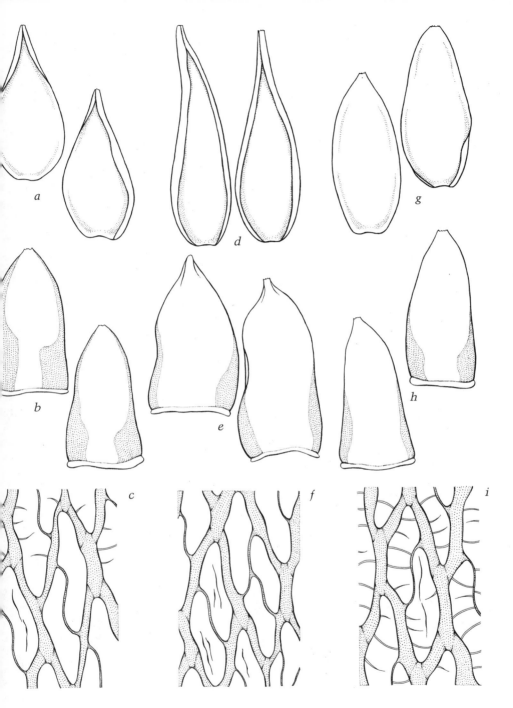

Fig. 8-107. *Sphagnum subfulvum* (a–c), *S. subnitens* (d–f), and *S. flavicomans* (g–i) compared as to stem and branch leaf shapes and upper cells of stem leaves, outer surface.

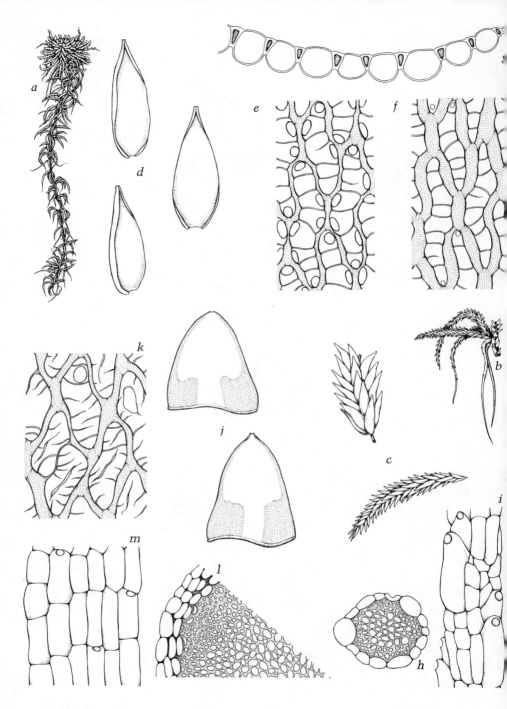

Fig. 8-108. *Sphagnum quinquefarium. a.* Habit. *b.* Fascicle of branches. *c.* Branches, enlarged. *d.* Branch leaves. *e.* Upper cells of branch leaf, outer surface. *f.* Upper cells of branch leaf, inner surface. *g.* Cross-section of branch leaf. *h.* Cross-section of branch. *i.* Branch epidermis. *j.* Stem leaves. *k.* Upper cells of stem leaf, outer surface. *l.* Cross-section of stem. *m.* Stem epidermis.

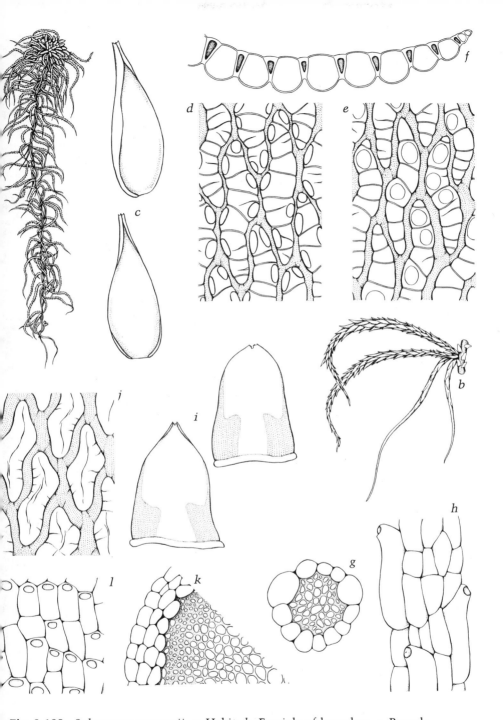

Fig. 8-109. *Sphagnum russowii.* a. Habit. b. Fascicle of branches. c. Branch leaves. d. Upper cells of branch leaf, outer surface. e. Upper cells of branch leaf, inner surface. f. Cross-section of branch leaf. g. Cross-section of branch. h. Branch epidermis. i. Stem leaves. j. Upper cells of stem leaf, outer surface. k. Cross-section of stem. l. Stem epidermis.

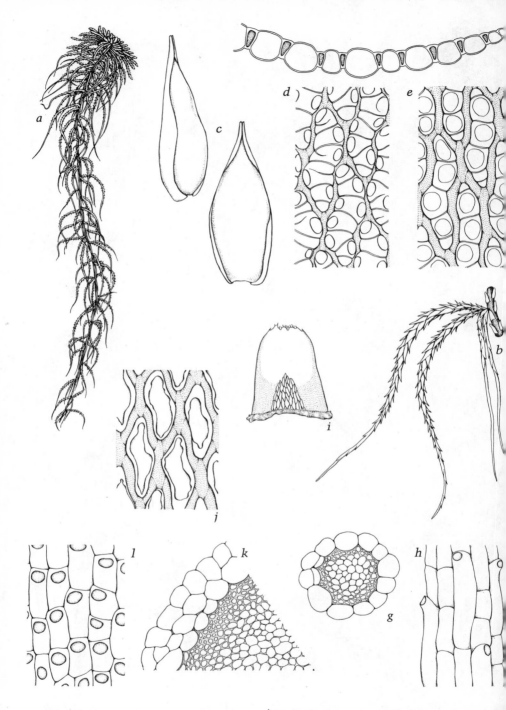

Fig. 8-110. *Sphagnum girgensohnii.* *a.* Habit. *b.* Fascicle of branches. *c.* Branch leaves. *d.* Upper cells of branch leaf, outer surface. *e.* Upper cells of branch leaf, inner surface. *f.* Cross-section of branch leaf, outer surface. *g.* Cross-section of branch. *h.* Branch epidermis. *i.* Stem leaf. *j.* Upper cells of stem leaf, inner surface (showing extensive resorption on both surfaces). *k.* Cross-section of stem. *l.* Stem epidermis.

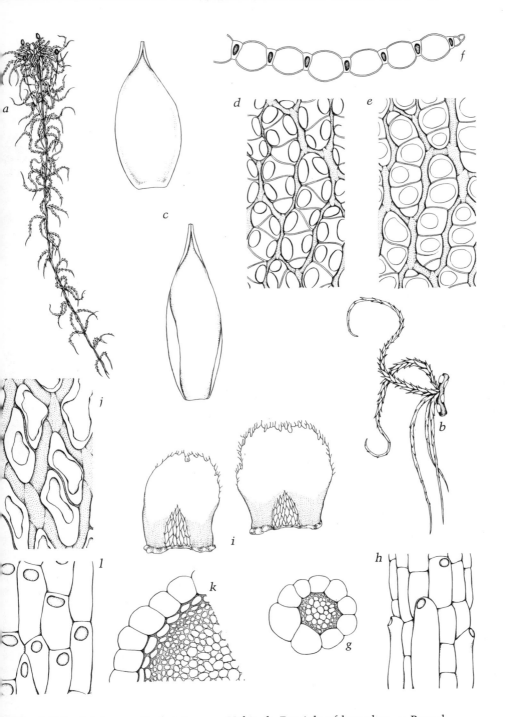

Fig. 8-111. *Sphagnum fimbriatum*. *a.* Habit. *b.* Fascicle of branches. *c.* Branch leaves. *d.* Upper cells of branch leaf, outer surface. *e.* Upper cells of branch leaf, inner surface. *f.* Cross-section of branch leaf. *g.* Cross-section of branch. *h.* Branch epidermis. *i.* Stem leaves. *j.* Upper cells of stem leaf, inner surface. *k.* Cross-section of stem. *l.* Stem epidermis.

Glossary

"When I use a word," Humpty Dumpty said in rather a scornful tone, "it means just what I choose it to mean—neither more nor less."
—Carroll

Terminology used in relation to the life history, structure, and taxonomy of *Sphagnum* is explained in the text, pp. 202–16.

aapamire A patterned peatland developed on a slight slope, with ridges of peat (*strings*) alternating with hollows or pools (*flarks*); also known as *string bog* or *string fen*

acidophile A plant adapted to acidic environments

airform patterns Appearance of peatlands of arctic and subarctic regions as viewed from altitudes of 30,000 feet, including *marbloid* (like polished marble), *terrazoid* (like patchwork), and *reticuloid* (like a network)

ammonification The microbial breakdown of amino acids resulting in the release of ammonia

apron An extensive downslope fen developed where the laggs on either side of a domed bog join

ash content The percentage of mineral solids remaining after a peat sample is burned

bog An ombrotrophic peatland, that is, one deriving water and nutrients only from the atmosphere; a highly acid and nutrient-poor peatland dominated by *Sphagnum* and ericaceous shrubs (and eventually, in North America, by black spruce), developed on peat elevated beyond the regional water table. *Basin bog*—occupying the basin of a pond or lake; a lake-fill bog. *Blanket bog*—covering irregular terrain, developed in oceanic areas with cool temperatures and persistently high humidity. *Continental bog*—an inland raised bog, less elevated and less convex than in oceanic regions. *Domed bog*—raised above ground level by a marked convexity, often with a concentric or eccentric pattern of ridges and depressions and/or pools. *Flat bog*—a fen developed under the influence of groundwater, lacking the convexity of a bog; also used in reference to any peatland that is flat or nearly so (including fens and inland raised bogs). *Marly bog*—rich fen, highly calcareous. *Plateau raised bog*—a relatively flat-topped, oceanic bog raised well above the level of the regional water table, with hummocks and hollows poorly differentiated and pools none or arranged irregularly (in contrast to domed bogs with subparallel patterning). *Quaking bog*—quagmire, floating mat, a wet fen marginal to a lake or pond, floating on water or on unconsolidated peat and yielding underfoot. *Raised bog*—any ombrotrophic peatland, or bog, but often used to denote oceanic bogs well

277

elevated above the regional water table and thus applied to domed and plateau raised bogs. Inland bogs, or *continental bogs*, are also raised but have only a slight elevation and convexity. *Slope bog*—occupying a sloping terrain but not covering the irregularities of topography that more extensive blanket bogs do; also known as a *hanging bog*. *String bog*—see *aapamire*, a less committal term which is more suitable because string bogs are essentially fens with some approach to bog conditions and vegetation on well-developed ridges of peat. *Transition bog*—a poor fen having a vegetation intermediate between a sedge mat and a bog; a wet, open, acid zone, dominated by *Sphagnum* but under the influence of ground or surface water.

bog iron Insoluble ferric (Fe^{+++}) iron compounds deposited under aerobic conditions in both bogs and fens

bog oak Wood of various kinds, including oak, especially remnants from ancient forests in the British Isles, buried by paludification beginning about 500 B.C.

boundary layer See *Grenzhorizont*

brown moss A true moss, as opposed to a "white moss," or peat moss

Btu British thermal unit, the amount of heat needed to raise the temperature of 1 pound of water 1° Fahrenheit

bulk density The weight per volume, or mass, of peat, including both organic and mineral content. The bulk density increases with decomposition.

calcareous soil water limit The demarcation between rich and poor fens or between sedge mats (intermediate fens) and *Sphagnum* lawns (poor fens)

capillary water Water occupying small pore spaces of the soil, moving in response to evaporation by means of cohesion of water molecules and adhesion to soil surfaces

carboxyl group The chemical radical -COOH held by polyuronic acids bound to the cell walls of *Sphagnum* (and other plants), involved in exchanging H^+ ions in exchange for cations of higher valence

carr A peatland containing broad-leaved trees and shrubs (such as red maple and poison sumac), characteristic of the outer margins of fen mats in the southern Great Lakes region

cation-exchange capacity The ability to release cations of low valence in exchange for those of higher valence. The high cation exchange by *Sphagnum* and *Sphagnum* peat depends on galacturonic acid in cell walls with its carboxyl groups, -COOH, capable of exchanging H^+ ions for metallic cations of higher valence.

chelation Chemical binding or complexing of metal cations (as contrasted with cation exchange, or adsorption), involving an equilibrium reaction between the ion and an organic molecule resulting in a stable ring structure

collapse scar A fenlike depression surrounded by a peat ridge, resulting from the melting of the ice core of a palsa or peat plateau

complexing See *chelation*

cupola The most elevated portion of a domed raised bog

denitrification The bacterial reduction of nitrates and nitrites to nitrous oxide or free nitrogen

drunken forest A stand of black spruce in subarctic regions of discontinuous permafrost, generally on peat plateaus where the ice core melts causing trees to lean or fall because of unstable peat in the substrate

dystrophic Rich in humic acids produced by a bog vegetation, used in reference to dark-stained, extremely oligotrophic bog waters

ericad A member of the Ericaceae, or heath family

eutrophic Rich in available nutrient ions. The word often has the connotation of high productivity owing to an abundance of phosphate and nitrate and also to an oxygen deficiency caused by an overpopulation of planktonic organisms.

false bottom A muddy suspension of undecayed organic matter, including planktonic remains, in minerotrophic bog lakes eventually exposed at the surface and allowing the advance of mat-forming sedges leading toward a fen-bog succession. (The term is sometimes used in reference to peat buoyed up by methane generation.)

fen Grass-, sedge-, or reed-dominated peatland, often with some shrub growth or a scant tree cover, developed under the influence of mineral-rich, aerated water at or near the surface. Fens with an abundance of calcium may develop into cedar swamps, or *treed fens*. Intermediate *fens* of lesser calcium-content may develop into *poor fens*, which are acid, through mineral-rich *Sphagnum* lawns and eventually into bogs.

fiber content The percentage of peat remaining after sieving under a gentle stream of water, giving an indication of the degree of decomposition. The *rubbed fiber* content provides a more meaningful index. It represents the percentage of organic matter that remains after firm rubbing 8–10 times between thumb and fingers.

fibric peat Scarcely decomposed peat consisting mainly of *Sphagnum* moss; see *humification scale*

flark See *aapamire*

floating mat See *quaking bog* under *bog*

fulvic acid See *humic substances*

Grenzhorizont The boundary layer of peat marking a change some 2,500 years ago toward cool, moist climates favorable to an increase in peat accumulation

grounded mat The bog stage of succession in a lake-fill peatland. The bog is built up beyond the level of the ground water because of peat accumulation. The grounded mat is joined at a hinge line to the floating mat made up of the pioneering fen stages (including intermediate sedge and poor fen stages).

gyttja Organic matter rich in nutrients deposited as grayish- or greenish-brown to blackish bottom mud made up of lake sediments such as marl, planktonic remains, and fecal material

hardpan An impermeable deposit of fine clay, humic acid, and oxides of iron and aluminum in the subsoil

heath A vegetation on well-drained mineral soil dominated by ericaceous shrubs. (With a significant ground cover of *Sphagnum*, such a heath may become a slope or blanket bog.) The word is also used to designate a shrubby member of the Ericaceae (or heath family).

hemic peat Peat of intermediate decomposition, reed-sedge peat; see *humification scale*

high-bush zone A shrubby zone of mineral enrichment marking the boundary between a bog and the surrounding vegetation, whether sloping woodland or swamp

hinge line Boundary between a floating fen and a grounded bog

humic substances Products of the incomplete decomposition of cellulose and lignin, consisting of humic and fulvic acids in addition to an ill-defined remnant called *humin* (fraction that is not dissolved on treatment with dilute alkali)

humification scale An indication of the degrees of decomposition of peat from *fibrous* (light-colored, scarcely decomposed *Sphagnum* peat) to *hemic* (partly decomposed peat derived from a reed-sedge vegetation) to *sapric* (dark, well-decomposed peat primarily derived from bottom sediments as well as sedge vegetation)

humin See *humic substances*

hummock A mound of *Sphagnum* commonly providing anchorage for ericaceous shrubs or bog trees, at maturity often topped by mosses and lichens. A *hummock-hollow complex* is the low-shrub stage of a bog, in which the hummocks are covered with a layer of *Chamaedaphne* and/or *Vaccinium*. It is often eventually occupied (in North America) by a black spruce muskeg.

hollow Wet depressions or pools among the hummocks that make up the bog surface

hydraulic conductivity The rate at which water moves through soil

ice-wedge polygon Many-sided area of patterned ground caused by permafrost, surrounded by drainage ditches filled with ice in winter, melt-water in summer

kettlehole A depression caused by melting of an ice block surrounded or covered over by till on glacial retreat. Bog lakes often occupy such depressions.

lagg The mineral-rich drainage area surrounding a bog, occupied by standing or sometimes moving water (as a *moat*), a sedgy fen, or a shrub growth

lake-fill The development of a mat marginal to a lake or pond and the eventual fill-in by an accumulation of peat; the process is also called *terrestrialization*

low-shrub zone The hummock-hollow, bog stage of succession consisting of mounds of *Sphagnum* covered over by ericaceous shrubs, especially *Chamaedaphne* and *Vaccinium*

macronutrients Ions needed in considerable quantity

marbloid See *airform pattern*

marl A deposit of calcium carbonate resulting from biotically induced changes in the carbonate-bicarbonate balance in freshwater basins, also as a result of evaporation or abrupt changes in temperature causing the escape of carbon dioxide from soluble calcium bicarbonate and the formation of insoluble calcium carbonate

marsh A grassy wetland developed on mineral soil in areas standing under water at least part of the year. Well aerated and rich in minerals, a marsh stores little or no peat. (A cattail marsh may develop on a peaty substrate at the mouth of an inlet stream and sometimes on a fen mat disturbed by foot traffic.)

mesotrophic With a mineral content intermediate between oligotrophic and eutrophic

micronutrients Trace elements needed for plant growth

minerals Usually referring to metallic cations in soil water solutions but also used more loosely to include anion nutrients such as phosphates and nitrates

mineral soil water limit The demarcation between minerotrophic and ombrotrophic conditions, that is, between fens and bogs

minerotrophic Receiving nutrients from the groundwater and therefore rich in minerals as compared to an ombrotrophic situation deriving minerals only from the atmosphere

mire Term for peatland used in noncommittal reference to fens and bogs. See also *aapamire*. Embryo mire—a small and incipient peatland of arctic or subarctic regions, developed especially in the ditches surrounding ice-wedge polygons. *Mesotrophic mire*—poor fen, an acid peatland receiving nutrients from ground water (as opposed to a bog). *Rheophilous mire*—peatland, or fen, under the influence of moving groundwater.

moor A peatland, any area of peat accumulation, whether minerotrophic or ombrotrophic. In England, the term refers to an upland vegetation consisting of ericaceous shrubs (such as heather).

moss Bog (rarely used as such in North American literature); also a bryophyte of the class Bryopsida, or true mosses, sometimes referred to as *brown mosses* in contrast to the Sphagnopsida, also known as *white mosses* or peat mosses. Brown mosses are abundant in rich fens.

muck Dark, well-decomposed peat high in ash content, largely derived from fen vegetation

muskeg An expanse of acid peatland bearing black spruce and an understory of ericaceous shrubs and a ground cover of *Sphagnum*. In Canada the term is commonly used synonymously with peatlands, wetlands accumulating one foot or more of peat.

mycorrhiza A root-fungus symbiosis important in the nutrient supply to plants, especially in poor soils. *Ectomycorrhizae* have roots enveloped in a mantle of fungal hyphae, and the hyphae lie among the cortical cells of the root but do not penetrate them. *Endomycorrhizae* have scarcely any mantle and the hyphae penetrate the cortical cells. *Ectendomycorrhizae* are intermediate in mantle development and have hyphae that penetrate the cortical cells.

nitrification The biological oxidation of ammonium and nitrite ions to the nitrate form

nitrogen fixation Reduction of free nitrogen to ammonia, especially by microorganisms, including certain bacteria, actinomycetes, and blue-green algae. Some nitrogen fixation is carried out by bacteria occupying root nodules in legumes or by actinomycetes in root nodules of *Alnus, Myrica*, and other shrubs.

nutrients Ions dissolved in soil water and taken up for use by plants

oligotrophic Poor in nutrients and therefore low in productivity

ombrotrophic Receiving nutrients exclusively from the atmosphere

paleoecology The study of past climates and vegetation based on an analysis of pollen, spores, and macrofossil remains deposited in sediments, including peat

palsa A peat-covered mound covering a permafrosted core. Palsas on coalescence form *peat plateaus* and *peat ridges*.

paludification Swamping, becoming wet by flooding, especially applied to the expansion of peatlands owing to a gradual rise in water table as peat accumulation impedes drainage

peat Organic matter deposited under water-soaked conditions as a result of incomplete decomposition. (*Fibric, hemic, sapric peat*—see *humification scale.*) *Peat moss*—*Sphagnum* moss; also used in reference to horticultural peat consisting mainly of *Sphagnum*. *Block-cut peat*—peat that is cut, whether by hand or by machine, into blocks and dried for handling. *Machine peat*—peat that is macerated and compressed by a hydraulic process. *Milled peat*—peat that is shredded and sieved. *Peat plateau*—extensive peat mounds formed by the coalescence of palsas. *Primary, secondary*, and *tertiary peat*—deposits corresponding to aquatic, sedge, and *Sphagnum* peat, associated with degrees of water retention and influence of groundwater.

peatland Any type of peat-covered terrain, including fens, bogs, and muskegs. *Patterned peatland*—see *aapamire*.

pingo Large mounds of alluvial soil covering an ice core formed in arctic lakes owing to hydraulic pressure forcing water through breaks in the permafrost and the buoyancy of ice formed as a result

podzolization The leaching of iron oxides into subsurface soil, associated with an acidic vegetation cover. Impermeable iron pans may result from podzolization.

von Post humification scale An index scale from 1 to 10 showing the degree to which peat is degraded; see *humification scale*

quaking mat See *quaking bog* (floating mat), under *bog*

rand The margin of a raised bog more or less abruptly sloped above the lagg; the slope between dome (or plateau) and lagg

recurrence surface A zone of abrupt change in the peat from dark and well decomposed to light-colored and fibrous, marking a change to cooler and moister conditions favorable to the development of bog vegetation and deposition of *Sphagnum* peat

reduction zone The anaerobic layers of peat where organic material accumulates owing to incomplete decomposition

reed-sedge peat Hemic peat composed of reeds, sedges, and grasses at the early (fen) stages of peatland successions, often laminated, stringy, and matted

regeneration complex An aggregation of hummocks and hollows on the bog surface, with bog growth achieved by the successive formations of hummocks on hollows and hollows on hummocks

reticuloid See *airform patterns*

rheophilous mire See *mire*

sapric peat Very decomposed peat; see *humification scale*

sedge meadow A sedge-dominated lowland similar to a fen but developed on mineral soil and less water-soaked during part of the season

slope bog (or *fen*)—See *bog*

soak A particularly wet area in a raised bog, often forming drainage channels, somewhat richer in minerals than the remainder of the surface

soligenous Referring to peatlands that receive water and nutrients from the surrounding soils

sphagnol A ligninlike component of the cell walls of *Sphagnum*

Sphagnum lawn Poor fen, a flat, wet, acid peatland, under the influence of groundwater, transitional to a bog

sporotrichosis A disease caused by *Sporothrix schenckii*, causing eruptions on the skin and eventually moving to vital organs through the lymphatic system, often contracted by handling horticultural peat, as well as rosebriers, splintered wood, etc.

string See *aapamire*

subsidence A lowering of the surface of agricultural peat soil because of microbial and oxidative decay, blowing, and shrinkage owing to drying

swamp A forested wetland, flooded during part of the year or with moving groundwater, well aerated, rich in minerals, and storing little or no peat. Swamps may be hardwood (with deciduous-leaved trees) or coniferous. Coniferous swamps developed on peat, such as *Thuja* swamps, are often called *treed fens*.

terrazoid See *airform patterns*
transition bog See *bog*
treed fen See *fen*, *swamp*

vernal pool A swamp depression filled with water at the spring highwater

water content The percentage of water by weight held in a peat sample, varying with the degree of decomposition

water table The top of the zone of saturation where all the pore spaces are filled with water in contrast to the aerated upper zone of peat

water track A peatland drainage area clogged with vegetation, somewhat richer in minerals than the rest of the boggy surface. In northern Minnesota, peatlands with strings and flarks are restricted to such drainage areas.

Bibliography

1. Ahern, P. J., and R. E. Bailey. 1980. Pollen record from Chippewa Bog, Lapeer County, Michigan. Mich. Academ. 12:297–308.
2. Aldrich, J. W. 1943. Biological survey of the bogs and swamps in northeastern Ohio. Amer. Midl. Nat. 30 (2): 346–402.
3. Allen, S. E., A. Carlisle, E. J. White, and C. C. Evans. 1968. The plant nutrient content of rainwater. Jour. Ecol. 56:497–504.
4. Ando, H., and A. Matsuo. 1984. Applied bryology. In Advances in Bryology, ed. W. Schultze-Motel, vol. 2, pp. 133–224 and 3 pls. Cramer, Vaduz.
5. Andrews, A. L. 1913. Sphagnales. No. Amer. Fl. 15:1–31.
6. Andrews, A. L. 1914–60. Notes on North American *Sphagnum*. Bryol. 14:72–75 (1911); 15:1–9, 63–66, 70–74 (1912); 16:20–24, 59–62, 74–76 (1913); 18:1–6 (1915); 20:84–89 (1917); 22:45–49 (1919); 24:81–86 ("1921," 1922); 61:269–76 ("1958," 1959); 62:87–96 (1959); 63:229–34 ("1960," 1961).
7. Andrews, A. L. 1937–51. Notes on the Warnstorf *Sphagnum* herbarium. Ann. Bryol. 9:3–12 ("1936," 1937); Bryol. 44:97–102, 155–59 (1941); 50:180–86 (1947); 52:124–30 (1949); 54:83–91 (1951).
8. Andrus, R. E. 1980. Sphagnaceae (peat moss family) of New York State. Contr. Fl. N.Y. State 3:1–89.
9. Andrus, R. E. 1986. Some aspects of *Sphagnum* ecology. Canad. Jour. Bot. 64:416–26.
10. Andrus, R. E., D. J. Wagner, and J. E. Titus. 1983. Vertical zonation of *Sphagnum* mosses along hummock-hollow gradients. Canad. Jour. Bot. 61:3128–39.
11. Auer, V. 1933. Peat bogs of southeastern Canada. Handbuch der Moorkunde 7:141–242.
12. Aulio, K. 1982. Nutrient accumulation in *Sphagnum* mosses. II. Intra- and interspecific variation in four species from ombrotrophic and minerotrophic habitats. Ann. Bot. Fenn. 19:93–101.
13. Bailey, R. E., and P. J. Ahern. 1981. A late- and post-glacial pollen record from Chippewa Bog, Lapeer County, Michigan: Further examination of white pine and beech immigration into the central Great Lakes region. In Geobotany II, ed. R. C. Romans, pp. 53–94. Plenum Press, New York.
14. Barber, K. E. 1981. Peat Stratigraphy and Climatic Changes: A Paleoecological Test of the Theory of Cyclic Peat Bog Regeneration. Balkema, Rotterdam.

15. Bartlein, P. J., and T. Webb III. 1982. Holocene climatic changes estimated from pollen data from the northern Midwest. *In* Quaternary History of the Driftless Area, pp. 67–82. Field Trip Guide, Book 5. Univ. of Wisconsin Extension Geological and Natural History Survey, Madison.
16. Basilier, K. 1980a. Moss-associated nitrogen fixation in some mire and coniferous forest environments around Uppsala, Sweden. Lindbergia 5:84–88. (1979).
17. Basilier, K. 1980b. Fixation and uptake of nitrogen in *Sphagnum* blue-green algal associations. Oikos 34:239–42.
18. Bellamy, D. J. 1968. An ecological approach to the classification of the lowland mires of Europe. *In* Proc. Third Intern. Peat Congr., Québec, National Research Council, Ottawa. pp. 74–79.
19. Birks, H. H. 1980. Plant microfossils in Quaternary lake sediments. Arch. Hydrobiol. 15:1–60.
20. Blomquist, H. L. 1938. Peatmosses of the southeastern states. Jour. Elisha Mitchell Sci. Soc. 54:1–21, pls. 1–5.
21. Boatman, D. K., and P. M. Lark. 1971. Inorganic nutrition of the protonema of *Sphagnum papillosum* Lindb., *S. magellanicum* Brid. and *S. cuspidatum.* Ehrh. New Phytol. 70:1053–59.
22. Boelter, D. H., and E. S. Verry, 1977. Peatland and water in the northern Lake States. Gen. Tech. Rept. NC-31. USDA For. Serv. St. Paul, Minn.
23. Brehm, K. 1971. Ein *Sphagnum*-Bult als Beispiel einer natürlichen Ionaustauschersäule. Beitr. Biol. Pfl. 47:287–312.
24. Brewer, R. 1966. The vegetation of two bogs in southwestern Michigan. Mich. Bot. 5:36–46.
25. Brown, D. H. 1982. Mineral nutrition. *In* Bryophyte Ecology, ed. A. J. E. Smith, pp. 383–444. Chapman and Hill, London and New York.
26. Buell, M. F., and H. F. Buell. 1941. Surface level fluctuations in Cedar Creek Bog, Minnesota. Ecol. 22:317–21.
27. Burgoff, H. 1954. Microbiologie des tourbières sur les resultats des recherches en commun nommées "Chiemseemoore und Bergener Becken." Huitième Internat. Bot. Paris, Rapp. et Comm. Sect. 7:154–56.
28. Catenhuisen, J. 1950. Secondary succession on the peatlands of Glacial Lake Wisconsin. Trans. Wisc. Acad. Sci. 40:29–48.
29. Cavers, F. 1911. The Inter-relationships of the Bryophyta. New Phytologist Reprint. Botany School, Cambridge.
30. Chapman, R. R., and H. F. Hemond. 1982. Dinitrogen fixation by surface peat and *Sphagnum* in an ombrotrophic bog. Canad. Jour. Bot. 60:538–43.
31. Clausen, J. J. 1957. A phytosociological ordination of the conifer swamps of Wisconsin. Ecol. 38:638–46.
32. Clymo, R. S. 1963. Ion-exchange in *Sphagnum* and its relation to bog ecology. Ann. Bot., n.s., 27:309–24.
33. Clymo, R. S. 1964. The origin of acidity in *Sphagnum* bogs. Bryol. 67:427–31.
34. Clymo, R. S. 1967. Control of cation concentrations, and particularly of pH, in *Sphagnum* dominated communities. *In* Chemical Environment in the Aquatic Habitat, ed. H. L. Golterman and R. S. Clymo, pp. 273–84. N. V. Noordhollandsche Uitgevers Maatachappij, Amsterdam.

35. Clymo, R. S. 1983. Peat. *In* Mires: Swamp, Bog, Fen and Moor, ed. A. J. P. Gore, pp. 159–224. Ecosystems of the World, vol. 4A. Elsevier, Amsterdam.
36. Clymo, R. S., and P. M. Hayward. 1982. The ecology of *Sphagnum*. *In* Bryophyte Ecology, ed. A. J. E. Smith, pp. 229–89. Chapman and Hill, London and New York.
37. Conway, V. M. 1949. The bogs of central Minnesota. Ecol. Monogr. 19:173–206.
38. Cooke, R. 1977. The Biology of Symbiotic Fungi. John Wiley, London.
39. Cooper, W. S. 1913. The climax forest of Isle Royale, Lake Superior, and its development. I. Bot. Gaz. 55:1–44.
40. Craigie, J. S., and W. S. G. Maass. 1966. The cation-exchanger in *Sphagnum* spp. Ann. Bot. 30:153–54.
41. Crow, G. E. 1969a. A phytogeographical analysis of a southern Michigan bog. Mich. Bot. 8:51–60.
42. Crow, G. E. 1969b. An ecological analysis of a southern Michigan bog. Mich. Bot. 8:11–27.
43. Crow, G. E. 1969c. Species of vascular plants of Pennfield Bog, Calhoun County, Michigan. Mich. Bot. 8:131–38.
44. Crum, H. 1983. Mosses of the Great Lakes Forest. 3d ed. Herbarium, Univ. of Michigan, Ann Arbor.
45. Crum, H. 1984. Sphagnopsida, Sphagnaceae. No. Amer. Fl., 2d ser., 11:1–180.
46. Crum, H. 1986. Sphagnaceae. *In* Illustrated Moss Flora of Arctic North America and Greenland, ed. G. S. Mogensen, pp. 1–61. Bioscience, vol. 18, pt. 2. Meddelelser om Grønland, Copenhagen.
47. Crum, H., and L. E. Anderson. 1981. Mosses of Eastern North America. 2 vols. Columbia Univ. Press, New York.
48. Curtis, J. T. 1959. Fen, meadow, and bog. *In* The Vegetation of Wisconsin, pp. 361–84. Univ. of Wisconsin Press, Madison.
49. Dachnowski-Stokes, A. P. 1933. Peat deposits in USA. Handbuch der Moorkunde 7:1–140.
50. Damman, A. W. H. 1965. Thin iron pans: Their occurrence and the conditions leading to their development. Info. Rept. N-X-2 Canad. For. Res. Cen., Sault Ste. Marie, Ontario.
51. Damman, A. W. H. 1977. Geographical changes in the vegetation pattern of raised bogs in the Bay of Fundy region of Maine and New Brunswick. Vegetatio 35:137–51.
52. Damman, A. W. H. 1978. Distribution and movement of elements in ombrotrophic peat bogs. Oikos 30:480–95.
53. Daniels, R. E., and A. Eddy. 1985. Handbook of European *Sphagna*. Institute of Terrestrial Ecology, Monks Wood Experimental Station, Huntington, England.
54. Davis, C. A. 1907. Origin and distribution of peat in Michigan. 1906 Rept. Mich. State Board Geol. Surv., Lansing, Mich.
55. Davis, C. A. 1911. The uses of peat for fuel and other purposes. U.S. Dept. Interior, Bur. Mines Bull. 16:1–214.
56. Davis, M. B. 1967. Late glacial climate in northern United States: A comparison of New England and the Great Lakes region. *In* Quaternary Paleoecology, ed. E. J. Cushing and H. E. Wright, Jr., pp. 11–43. Yale Univ. Press, New Haven.
57. Davis, M. B. 1983. Holocene vegetational history of the eastern United States. *In*

Late-Quaternary Environment of the United States, ed. H. E. Wright, Jr., pp. 166–81. Vol. 2, The Holocene. Univ. of Minnesota Press, Minneapolis.
58. Deevey, E. S., Jr. 1942. Studies in Connecticut lake sediments. III. The biostratonomy of Linsley Pond. Amer. Jour. Sci. 240:233–64, 313–24.
59. Deevey, E. S., Jr. 1958. Bogs. Sci. Amer., Oct., pp. 1–8.
60. Delcourt, P. A. and H. R. Delcourt. 1981. Vegetation maps for eastern North America: 40,000 yr. B.P. to the present. In Geobotany II, ed. R. C. Romans, pp. 123–65. Plenum Press, New York.
61. Dice, L. R. 1932. A preliminary classification of the major terrestrial ecological communities of Michigan exclusive of Isle Royale. Pap. Mich. Acad. 16:217–39.
62. Dickinson, W. 1975. Recurrence surfaces in Rusland Moss, Cambria (formerly North Lancashire). Jour. Ecol. 63:913–35.
63. Dorr, J. A., Jr., and D. F. Eschman. 1970. Geology of Michigan. Univ. of Michigan Press, Ann Arbor.
64. Drlica, K. 1982. A burning question. Garden, Nov./Dec., pp. 26–29.
65. DuRietz, G. E. 1954. Die Mineralbodenwasserzeigergrenze als Grundlage einer natürlichen Zweigliederung der nord- und mitteleuropäischen Moore. Vegetatio 5–6: 571–85.
66. Eddy, A. 1977. Sphagnales of tropical Asia. Bull. Brit. Mus. Bot. 5:359–446.
67. Eddy, A. 1979. Taxonomy and evolution of *Sphagnum*. In Bryophyte Systematics, ed. G. C. S. Clarke and J. G. Duckett, pp. 109–21. Academic Press, London and New York.
68. Eurola, S., and A. Huttunen, eds. 1985. Proceedings of the field symposium on classification of mire vegetation, Hailusto-Kuusamo, Sept. 5–13, 1983. Aquilo, bot. ser., 21:1–121.
69. Farnham, R. S., and D. N. Grubich. 1966. Peat resources of Minnesota. Rept. Inventory no. 3, Red Lake Bog, Beltrami County, Minnesota. State of Minnesota Office of Iron Range Resources and Rehabilitation, St. Paul.
70. Field, E. M., and D. A. Goode. 1981. Peatland Ecology in the British Isles: A Bibliography. Institute of Terrestrial Ecology and Nature Conservancy Council, London.
71. Foster, D. R., G. A. King, P. H. Glaser, and H. E. Wright, Jr. 1983. Origin of string patterns in boreal peatlands. Nature 306:256–58.
72. Fuchsman, C. H. 1980. Peat: Industrial Chemistry and Technology. Academic Press, New York.
73. Futyma, R. P. 1982. Postglacial vegetation of eastern Upper Michigan. Ph.D. diss., Univ. of Michigan.
74. Futyma, R. P., and N. G. Miller. 1987. Stratigraphy and genesis of the Lake Sixteen peatland, northern Michigan. Canad. Jour. Bot. 64:3008–19.
75. Gates, F. C. 1912. The vegetation of the region in the vicinity of Douglas Lake, Cheboyan County, Michigan, 1911. Rept. Mich. Acad. 14:46–106.
76. Gates, F. C. 1914. Winter as a factor in the xerophily of certain evergreen ericads. Bot. Gaz. 57:445–89.
77. Gates, F. C. 1917. The relation between evaporation and plant succession in a given area. Amer. Jour. Bot. 4:161–78.

78. Gates, F. C. 1926a. Evaporation in vegetation at different heights. Amer. Jour. Bot. 13:167–78.
79. Gates, F. C. 1926b. Plant successions about Douglas Lake, Cheboygan County, Michigan. Bot. Gaz. 82:170–82.
80. Gates, F. C. 1940. Bog levels. Science 91:449–50.
81. Gates, F. C. 1942a. Plant succession. Kans. Acad. Sci. Trans. 45:27–35.
82. Gates, F. C. 1942b. The bogs of northern lower Michigan. Ecol. Monogr. 12:213–54.
83. Gauthier, R. 1980. La Végétation des tourbières et les sphaignes du parc des Laurentides. Université Laval, Québec.
84. Gauthier, R., and M. M. Grandtner. 1975. Etude phytosociologique des tourbières du Bas Saint-Laurent, Québec. Nat. Canad. 102:109–53.
85. Glaser, P. H. 1983a. A patterned fen on the north shore of Lake Superior, Minnesota. Canad. Field-Natur. 97:194–99.
86. Glaser, P. H. 1983b. Vegetation patterns in the North Black River peatland, northern Minnesota. Canad. Jour. Bot. 61:2085–2104.
87. Glaser, P. H., and G. A. Wheeler. 1980. The development of surface patterns in Red Lake Peatland, northern Minnesota. In Proc. Sixth Intern. Peat Congr., Duluth, pp. 31–45. International Peat Society, Duluth.
88. Glaser, P. H., G. A. Wheeler, E. Gorham, and H. E. Wright, Jr. 1981. The patterned mires of the Red Lake Peatland, northern Minnesota: Vegetation, water chemistry, and landforms. Jour. Ecol. 69:575–99.
89. Glime, J. M., R. G. Wetzel, and B. J. Kennedy. 1982. The effects of bryophytes on succession from alkaline marsh to *Sphagnum* bog. Amer. Midl. Natur. 108:209–23.
90. Glob, P. V. 1970. The Bog People: Iron Age Man Preserved. Cornell Univ. Press, Ithaca.
91. Godwin, H. 1978. Fenland: Its Ancient Past and Uncertain Future. Cambridge Univ. Press, Cambridge.
92. Godwin, H. 1981. The Archives of the Peat Bogs. Cambridge Univ. Press, Cambridge.
93. Goff, F. G., J. P. Ludwig, S. P. Voice, and H. A. Tanner. 1983. An assessment of the impacts expected from the proposed *Sphagnum* Peat Mining Project by Garrett Peat Industries, Inc., in the Dingman Basin, Hebron Township, Cheboygan County, Michigan. Spec. Rept. DMP-83-3. Ecological Research Services, Inc., Boyne City, Mich.
94. Gore, A. J. P., ed. 1983. Mires: Swamp, Bog, Fen and Moor. Ecosystems of the World, vols. 4A–B. Elsevier, Amsterdam.
95. Gorham, E. 1956. The ionic composition of some bog and fen waters in the English Lake District. Jour. Ecol. 44:142–52.
96. Gorham, E. 1957. The development of peatlands. Quart. Rev. Biol. 32:145–66.
97. Gorham, E. 1961. Factors influencing supply of major ions to inland waters, with special references to the atmosphere. Bull. Geol. Soc. Amer. 72:795–840.
98. Gorham, E., and W. H. Pearsall. 1956. Acidity, specific conductivity and calcium content of some bog and fen waters in northern Britain. Jour. Ecol. 51:928–30.

99. Hansen, B. 1961–66. Sphagnaceae. Dansk Bot. Ark. 20:89–108 (1961), 204 (1962); 23:295–300 (1966).
100. Harshberger, J. W. 1909. Bogs, their nature and origin. Plant World 12:32–41, 53–61.
101. Heinselman, M. L. 1963. Forest sites, bog processes, and peatland types in Glacial Lake Agassiz region. Ecol. Monogr. 33:327–74.
102. Heinselman, M. L. 1965. String bogs and other patterned organic terrain near Seney, upper Michigan. Ecol. 46:185–88.
103. Heinselman, M. L. 1970. Landscape evolution, peatland types, and the environment in the Lake Agassiz Peatlands Natural Area, Minnesota. Ecol. Monogr. 40:235–61.
104. Heinselman, M. L. 1975. Boreal peatlands in relation to environment. In Ecological Studies, ed. A. D. Hasler, pp. 93–103. Vol. 10, Coupling of Land and Water Surfaces. Springer, New York.
105. Hemond, H. F. 1980. Biogeochemistry of Thoreau's bog, Concord, Massachusetts. Ecol. Monogr. 50:507–26.
106. Hemond, H. F. 1983. The nitrogen budget of Thoreau's bog. Ecol. 64:99–109.
107. Hill, M. O. 1976. A key for the identification of British *Sphagnum* using macroscopic characters. Bull. Brit. Bryol. Soc. 27:22–31.
108. Hill, M. O. 1977. A critical assessment of the distinction between *Sphagnum capillaceum* (Weiss) Schrank and *S. rubellum* Wils. in Britain. Bull. Brit. Bryol. Soc. 29:19.
109. Hill, M. O. 1978. Sphagnopsida. In The Moss Flora of Britain and Ireland, ed. A. J. E. Smith, pp. 30–78. Cambridge Univ. Press, Cambridge.
110. Hills, J. L., and F. M. Hollister. 1912. The peat and muck deposits in Vermont. Vermont Agric. Exper. Sta. Bull. 165:141–240.
111. Holcombe, J. W. 1984. Morphogenesis of branch leaves of *Sphagnum magellanicum* Brid. Jour. Hattori Bot. Lab. 57:179–240.
112. Holmen, H., A. Johnels, N. Malmer, Å. Persson, and H. Sjörs. 1967. Peatland and peatland conservation in Sweden. Aquilo, bot. ser., 6:120–36.
113. Hörmann, H. 1965. Vom Wasserhaushalt der Moose. Mikrokosmos 8:245–49.
114. Hutchinson, G. E., et al. 1970. Ianula: An account of the Lago di Monterosi, Latium, Italy. Trans. Amer. Phil. Soc., n.s., 60:1–178.
115. Isoviita, P. 1966. Studies on *Sphagnum* L. I. Nomenclatural revision of the European taxa. Ann. Bot. Fenn. 3:199–264.
116. Jeglum, J. K. 1971. Plant indicators of pH and water level in peatlands at Candle Lake, Saskatchewan. Canad. Jour. Bot. 49:1661–76.
117. Jeglum, J. K., A. N. Boissonneau, and V. F. Haavisto. 1974. Toward a wetland classification for Ontario. Info. Rept. 0-X-215. Canad. For. Serv., Sault Ste. Marie, Ontario.
118. Johnson, C. W. 1985. Bogs of the Northeast. Univ. Press of New England, Hanover and London.
119. Joyal, R. 1970. Description de la tourbière à sphaignes Mer Bleue près d'Ottawa. I. Végétation. Canad. Jour. Bot. 48:1405–18.
120. Joyal, R. 1971. The Marshall Bog near Itasca Park, Minnesota. Mich. Bot. 10:78–88.

121. Joyal, R. 1972. La tourbière à sphaignes Mer Bleue près d'Ottawa. II. Quelques facteurs ecologique. Canad. Jour. Bot. 50:1209–18.
122. Judd, W. W. 1957. Studies of the Byron Bog in southwestern Ontario. Pt. 1 Description of the bog. Canad. Entom. 89:235–38.
123. Kapp, R. O. 1969. How to Know Pollen and Spores. Wm. C. Brown, Dubuque.
124. Kapp, R. O. 1986. Late glacial pollen and macrofossils associated with the Rappuhn mastodont (Lapeer County, Michigan). Amer. Midl. Nat. 116 (2): 368–77.
125. Katz, N. J., S. V. Katz, and E. I. Skobeeva. 1977. Atlas of Plant Remains in Peat Deposits [in Russian]. "Ne-ra," Moscow.
126. Ketchledge, E. H. 1964. The ecology of a bog. N.Y. State Conserv., June–July 1964, unpaged.
127. Kilham, P. 1982a. Acid precipitation: Its rôle in the alkalization of a lake in Michigan. Limnol. & Oceanogr. 27:856–67.
128. Kilham, P. 1982b. The biogeochemistry of bog ecosystems and the chemical ecology of *Sphagnum*. Mich. Bot. 21:159–68.
129. Kulczynski, S. 1949. Peat bogs of Polesie. Mem. Acad. Sci. Cracovie, ser. B, 15:1–356.
130. Lange, B. 1982. Key to northern boreal and arctic species of *Sphagnum* based on characteristics of the stem leaves. Lindbergia 8:1–29.
131. Larsen, J. A. 1982. Ecology of the Northern Lowland Bogs and Conifer Forests. Academic Press, New York.
132. Levathes, L. E. 1987. Mysteries of the bog. Amer. Geogr. 171 (3): 397–420.
133. Lindsay, R. 1987. The Great Flow—an international responsibility. New Scientist, Jan. 8, p. 45.
134. Lindstrom, O. 1980. The technology of peat. Ambio, Jour. Human Environ. 9:309–13.
135. Looman, J. 1982. The vegetation of the Canadian prairie provinces. III. Aquatic and semi-aquatic vegetation, pt. 2. Freshwater marshes and bogs. Phytocoenologia 10:401–23.
136. Lucas, R. E. 1982. Organic Soils (histosols): Formation, distribution, physical and chemical properties and management for crop production. Farm Sci. Research Rept. 435. Michigan State Agricultural Experiment Station, East Lansing.
137. Lucas, R. E., and J. F. Davis. 1961. Relationships between pH values of organic soils and availabilities of 12 plant nutrients. Soil Sci. 92:177–82.
138. Ludwig, J. P., F. G. Goff, and S. P. Voice. 1983. A response to a request for proposal and data on the Dingman Peatland. Ecological Research Services, Inc., Iron River, Mich.
139. Luken, J. D., and W. D. Billings. 1986. Hummock-dwelling ants and the cycling of microtopography in an Alaskan peatland. Canad. Field-Natur. 100:69–73.
140. Madsen, B. J. 1987. Interaction of vegetation and physical processes in patterned peatlands: A comparison of two sites in upper Michigan. Ph.D. diss., Univ. of Michigan.
141. Mägdefrau, K. 1935. Untersuchungen über die Wasserversorgung der Gametophyten und Sporophyten der Laubmoose. Zeitschr. Bot. 29:337–75.

142. Malmer, N. 1958. Notes on the relation between the chemical composition of mire plants and peat. Bot. Nat. 3:274–88.
143. Malmer, N., and H. Sjörs. 1955. Some determinations of elementary constituents in mire plants and peat. Bot. Nat. 108:46–80.
144. Martin, N. D. 1959. An analysis of forest succession in Algonquin Park, Ontario. Ecol. Monogr. 29:187–218.
145. McAndrews, J. H., and A. A. Berti. 1973. Key to Quaternary Pollen and Spores of the Great Lakes Region. Life Sci. Misc. Publ. 100, pp. 1–61. Royal Ontario Museum, Queen's Park, Toronto.
146. Meades, W. J. 1983. Heathlands. In Biogeography and Ecology of the Island of Newfoundland, ed. G. R. South, pp. 267–318. Junk, The Hague.
147. Miller, N. G. 1981. Bogs, bales, and Btu's: A Primer on peat. Horticulture 59:38–45.
148. Miller, N. G., and R. P. Futyma. 1987. Paleohydrological implications of Holocene peatland development in northern Michigan. Quat. Research 27:297–311.
149. Moore, J. J. 1968. A classification of the bogs and wet heaths of northern Europe. In Berichte Intern. Symp. pflanzensoz. Syst., ed. R. Tüxen, pp. 300–320. Gustav Fischer, Jena.
150. Moore, P. D. 1975. Origin of blanket mires. Nature 256:267–69.
151. Moore, P. D. 1987. A thousand years of death. New Scientist, Jan. 8, pp. 46–48.
152. Moore, P. D., and D. J. Bellamy. 1973. Peatlands. Springer, New York.
153. Moore, P. D., and J. A. Webb. 1978. An Illustrated Guide to Pollen Analysis. Halsted Press, New York.
154. Morgan, S. M., and F. C. Pollett. 1983. Proceedings of a Peatland Inventory Methodology Workshop. Land Resource Research Institute, Agriculture Canada, Ottawa.
155. Moss, E. M. 1953. Marsh and bog vegetation in northwestern Alberta. Canad. Jour. Bot. 31:448–70.
156. Newbould, P. J. 1958. Peat bogs. New Biology 26:88–105.
157. Nichols, G. E. 1915. The vegetation of Connecticut. IV. Plant societies in lowlands. Bull. Torry Bot. Club 42:169–217.
158. Nichols, G. E. 1918a. The *Sphagnum* moss and its use in surgical dressings. Jour. N.Y. Bot. Gard. 19:203–20, pls. 216–18.
159. Nichols, G. E. 1918b. The vegetation of northern Cape Breton Island, Nova Scotia. Trans. Conn. Acad. Arts & Sci. 22:249–467.
160. Nichols, G. E. 1918c. War work for bryologists. Bryol. 21:53–56.
161. Nichols, G. E. 1919. Raised bogs of eastern Maine. Geogr. Rev. 7:159–67.
162. Nichols, G. E. 1920. Sphagnum moss: War substitute for cotton in absorbent surgical dressings. Smiths. Rept. 1918:221–34, pls. 1–4.
163. Noble, M. G., D. B. Lawrence, and G. P. Streveler. 1984. *Sphagnum* invasion beneath an evergreen forest canopy in southeastern Alaska. Bryol. 87:119–27.
164. Noguchi, A. 1958. Germination of spores in two species of *Sphagnum*. Jour. Hattori Bot. Lab. 19:71–75.
165. Nyholm, E. 1969. Sphagnales. In Illustrated Moss Flora of Scandinavia, pp. 697–765. II. Musci, fasc. 6. Swedish Natural Science Research Council, Stockholm.

166. Osvald, H. 1923. Die Vegetation des Hochmoores-Komosse. Sv. Växtsoc. Sällsk. Handb. 1:1–436.
167. Osvald, H. 1925. Die Hochmoortypen Europas. Veröff. Geobot. Inst. Rübel Zürich 3:707–23.
168. Osvald, H. 1928. Nordamerikanska moostyper. Sv. Bot. Tidskr. 22:377–91.
169. Osvald, H. 1935. A bog at Hartford, Michigan. Ecol. 16:520–28.
170. Osvald, H. 1940. *Sphagnum flavicomans* (Card.) Warnst.: Taxonomy and ecology. Acta Phytogeogr. Suec. 13:39–49, pls. 1, 4.
171. Osvald, H. 1949. Notes on the vegetation of British and Irish mosses. Acta Phytogeogr. Suec. 26:1–62, pls. 1–16.
172. Osvald, H. 1955. The vegetation of two raised bogs in northeastern Maine. Sv. Bot. Tidskr. 49:110–18.
173. Osvald, H. 1970. Vegetation and stratigraphy of peatlands in North America. Acta Univ. Upsal., ser. 5C, 1:1–96.
174. O'Toole, M. D., and O. M. Synnott. 1971. The bryophyte succession on blanket peat following calcium carbonate, nitrogen, phosphorus, and potassium fertilizers. Jour. Ecol. 59:121–26.
175. Overbeck, F., and H. Happach. 1957. Über das Wachstum und den Wasserhaushalt einiger Hochmoor-Sphagnen. Flora 144:335–402.
176. Pakarinen, P. 1977. Element content of *Sphagnum:* Variation and sources. Bryoph. Biblio. 13:751–62.
177. Pakarinen, P. 1978. Production and nutrient ecology of three *Sphagnum* species in southern Finnish raised bogs. Ann. Bot. Fenn. 15:15–26.
178. Pakarinen, P. 1979. Ecological indicators and species groups of bryophytes in boreal peatlands. *In* Classification of Peat and Peatlands, pp. 121–34. Proc. Intern. Symp. Hyytiälä, Finland, Sept. 17–21, 1979.
179. Pakarinen, P., and K. Tolonen. 1977a. Nutrient contents of *Sphagnum* mosses in relation to bog water chemistry in northern Finland. Lindbergia 4:27–33.
180. Pakarinen, P., and K. Tolonen. 1977b. Vertical distributions of N, P, K, Zn and Pb in *Sphagnum* peat. Suo 28 (4–5): 95–102.
181. Paul, H. 1924. Sphagnaceae (Torfmoose). *In* Die natürlichen Pflanzenfamilien, ed. A. Engler and K. Prantl, pp. 105–25. 2d ed. Vol. 10. Engelmann, Leipzig.
182. Paul, H. 1932. Der Einfluss des Wassers auf die Gestaltungsverhältnisse der *Sphagna.* Abh. Natür.-Ver. Bremen 28:78–96.
183. Pearsall, W. H. 1941. The "mosses" of the Stainmore District. Jour. Ecol. 29:161–75.
184. Pearsall, W. H. 1950. Mountains and Moorlands. New Naturalist, ser. 11. Collins, London.
185. Pearsall, W. H., and E. M. Lind. 1941. A note on a Connemara bog type. Jour. Ecol. 29:62–68.
186. Pollett, F. C. 1968. Peat resources of Newfoundland. Mineral Resources Rept. 2. Prov. of Newfoundland and Labrador Dept. of Mines, Agric. & Resources, St. John's, Newfoundland.
187. Pollett, F. C., A. F. Rayment, and A. Robertson, eds. 1979. The Diversity of Peat. Newfoundland and Labrador Peat Association, St. John's.

188. von Post, L. 1922. Sveriges geologiska undersökningens torvinventering ash nagra av dess hittills vunna resultat. Sv. Mosskulturför. 1:1–27.
189. von Post, L., and R. Sernander. 1910. Pflanzenphysiognomische Studien auf Torfmooren in Närke. Livret. Excurs. Suèd. xi Congr. Geol. Intern. 1 (14): 1–48.
190. Potzger, J. E. 1942a. Forest succession in the Trout Lake, Vilas County, Wisconsin area: A pollen study. Butler Univ. Bot. Stud. 5:179–89.
191. Potzger, J. E. 1942b. Pollen spectra from four bogs on the Gillen Nature Reserve, along the Michigan-Wisconsin state line. Amer. Midl. Nat. 28 (2): 501–11.
192. Potzger, J. E. 1943. Pollen study of five bogs in Price and Sawyer counties, Wisconsin. Butler Univ. Bot. Stud. 6:54–64.
193. Potzger, J. E. 1948. A pollen study in the tension zone of lower Michigan. Butler Univ. Bot. Stud. 8:161–77.
194. Potzger, J. E. 1950. Bogs of the Quetico-Superior Country Tell Its Forest History. President's Quetico-Superior Committee, AAAS, Urbana, Ill.
195. Potzger, J. E. 1951. The fossil record near the glacial border. Ohio Jour. Sci. 51:126–33.
196. Potzger, J. E. 1953. History of forests in the Quetico-Superior Country from fossil pollen studies. Jour. Forestry 51:560–65.
197. Potzger, J. E., and A. Courtemanche. 1955. Permafrost and some characteristic bogs and vegetation of northern Quebec. Rev. Canad. Géogr. 9:109–14.
198. Potzger, J. E., and A. Courtemanche. 1956. A series of bogs across Quebec from the St. Lawrence Valley to James Bay. Canad. Jour. Bot. 34:473–500.
199. Potzger, J. E., and R. C. Friesner. 1948. Forests of the past along the coast of southern Maine. Butler Univ. Bot. Stud. 8:178–203.
200. Potzger, J. E., and J. H. Otto. 1943. Post-glacial forest succession in northern New Jersey as shown by pollen records from five bogs. Amer. Jour. Bot. 30:83–87.
201. Press, M. C., and J. A. Lee. 1982. Nitrate reductase activity of *Sphagnum* species in the South Pennines. New Phytol. 92:487–94.
202. Pringle, J. S. 1980. An introduction to wetland classification in the Great Lakes region. Techn. Bull. Roy. Bot. Gard. (Hamilton, Ont.) 10:1–11.
203. Puustjärvi, V. 1952. The precipitation of iron in peat soils. Acta Agralia Fenn. 78:1–72.
204. Radforth, N. W., and C. O. Brawner. 1977. Muskeg and the Northern Environment in Canada. Univ. of Toronto Press.
205. Ramant, J. 1955. Étude de l'origine de l'acidité naturelle des tourbières acides de la Boraque-Michel. Bull. Acad. Roy. Belg. Sci., 5th ser., 41: 1037–52.
206. Ratcliffe, D. A., and D. Walker. 1958. The Silver Flowe, Galloway, Scotland. Jour. Ecol. 46:407–45.
207. Rawson, D. S. 1956. Algal indicators of trophic lake types. Limnol. & Oceanogr. 1:18–25.
208. Redhead, S. A., and K. W. Spicer. 1981. *Discinella schimperi*, a circumpolar parasite of *Sphagnum squarrosum*, and notes on *Bryophytomyces sphagni*. Mycologia 73:904–13.
209. Rigg, G. B. 1940–51. The development of *Sphagnum* bogs in North America. Bot. Rev. 6:666–93 (1940); 17:109–31 (1951).

210. Roberts, B. A., and A. W. Robertson. 1980. Palsa bogs, sand dunes and salt marshes, environmentally sensitive habitats in the coastal region, southeastern Labrador. In Workshop on Research on the Labrador Coastal and Offshore Region, pp. 245–63. Newfoundland Inst. for Cold Water Research, Memorial Univ., St. John's.
211. Rose, F. 1953. A survey of the ecology of the British lowland bogs. Proc. Linn. Soc. London 164:186–211.
212. Rudolph, H., and N. Jöhnk. 1982. Physiological aspects of phenolic compounds in the cell walls of Sphagna. Jour. Hattori Bot. Lab. 53:195–203.
213. Ruhland, W. 1924. Musci. Allgemeiner Teil. In Die natürlichen Pflanzenfamilien, ed. A. Engler and K. Prantl, pp. 1–105. 2d ed. Vol. 10. Engelmann, Leipzig.
214. Ruuhijärvi, R. 1983. The Finnish mire types and their regional distribution. In Mires: Swamp, Bog, Fen and Moor, ed. A. J. P. Gore, pp. 47–67. Ecosystems of the World, vol. 4B. Elsevier, Amsterdam.
215. Rymal, D. E., and G. W. Folkerts. 1982. Insects associated with pitcher plants (Sarracenia: Sarraceniaceae) and their relationship to pitcher plant conservation: A review. Jour. Alabama Acad. Sci. 53:131–51.
216. Sanger, R., and J. E. Gannon. 1979. Vegetation succession in Smith's Bog, Cheboygan County, Michigan. Mich. Bot. 18:59–69.
217. Schimper, W. P. 1857. Mémoire pour servir à l'histoire naturelle des sphaignes (Sphagnum L.). Mém. Acad. Sci. Sav. Inst. Fr. 15:1–96, pls. 1–24.
218. Schimper, W. P. 1858. Versuch einer Entwicklungsgeschichte der Torfmoose (Sphagnum) und einer Monographie der in Europa vorkommenden Arten dieser Gattung. Schweizerbart, Stuttgart.
219. Schwintzer, C. R. 1975. Vegetation and nutrient status of northern Michigan fens. Canad. Jour. Bot. 56:3044–51.
220. Schwintzer, C. R. 1978. Nutrients and water levels in a small Michigan bog with high tree mortality. Amer. Midl. Nat. 100 (2): 441–51.
221. Schwintzer, C. R. 1979. Vegetation changes following a water level rise and tree mortality in a Michigan bog. Mich. Bot. 18:91–98.
222. Schwintzer, C. R. 1983. Nonsymbiotic and symbiotic nitrogen fixation in a weakly minerotrophic peatland. Amer. Jour. Bot. 70:1071–78.
223. Schwintzer, C. R., and J. R. Tomberlin. 1982. Chemical and physical characteristics of shallow groundwaters in northern Michigan bogs, swamps, and fens. Amer. Jour. Bot. 69:1231–39.
224. Schwintzer, C. R., and G. Williams. 1974. Vegetation changes in a small Michigan bog from 1917 to 1972. Amer. Midl. Nat. 92 (2): 447–59.
225. Sears, P. B. 1930. A record of post-glacial climate in northern Ohio. Ohio Jour. Sci. 30:205–17.
226. Sears, P. B. 1932. Post-glacial climate in eastern North America. Ecol. 13:1–6.
227. Sears, P. B. 1935. Glacial and post-glacial vegetation. Bot. Rev. 1:37–51.
228. Siegel, D. I. 1983. Groundwater and the evolution of patterned mires, glacial Lake Agassiz peatlands, northern Minnesota. Jour. Ecol. 71:913–21.
229. Sjörs, H. 1948. Myrvegetation in Bergslagen. Acta Phytogeogr. Suec. 21:1–299, pls. 1–32.

230. Sjörs, H. 1950a. On the relation between vegetation and electrolytes in north Swedish mires. Oikos 2:243–58.
231. Sjörs, H. 1950b. Regional studies in north Swedish mire vegetation. Bot. Not. 2:173–222.
232. Sjörs, H. 1959. Bogs and fens in the Hudson Bay Lowlands. Arctic 12:1–19.
233. Sjörs, H. 1961a. Forest and peatland at Hawley Lake, northern Ontario. Bull. Nat. Mus. Canada 171:1–31, pls. 1–5.
234. Sjörs, H. 1961b. Surface patterns in boreal peatland. Endeavour 20:217–24.
235. Sjörs, H. 1963. Bogs and fens on Attawapiskat River, northern Ontario. Bull. Nat. Mus. Canada 186:45–131, pls. 1–27.
236. Sjörs, H. 1980. Peat on earth: Multiple use or conservation? Ambio, Jour. Human Environ. 9:303–8.
237. Sjörs, H. 1983. Mires of Sweden. In Mires: Swamp, Bog, Fen and Moor, ed. A. J. P. Gore, pp. 69–94. Ecosystems of the World, vol. 4B. Elsevier, Amsterdam.
238. Soper, E. K. 1919. The peat deposits of Minnesota. Minn. Geol. Surv. Bull. 16:1–261.
239. Sorenson, E. 1986. The Ecology and Distribution of Patterned Fens in Maine. Maine State Planning Office, Augusta.
240. Spearing, A. M. 1972. Cation-exchange capacity and galacturonic acid content of several species of *Sphagnum* in Sandy Ridge Bog, central New York State. Bryol. 75:154–58.
241. Spurr, S. H. 1956. Michigan's forests over ten thousand years. Mich. Alum. Quart. Rev. 62:336–41.
242. Stocker, O. 1956a. Die Wasserleitung der Thallophyten. In Handbuch der Pflanzenphysiologie, ed. W. Ruhland, pp. 514–21. Vol. 3. Springer, Berlin.
243. Stocker, O. 1956b. Wasser und Wasserspeicherung bei Thallophyten. In Handbuch der Pflanzenphysiologie, ed. W. Ruhland, pp. 161–72. Vol. 3. Springer, Berlin.
244. Suzuki, H. 1972. Distribution of *Sphagnum* species in Japan and an attempt to classify the moors basing on their combination. Jour. Hattori Bot. Lab. 35:3–24.
245. Tansley, A. G. 1939. The British Islands and Their Vegetation. Cambridge Univ. Press, Cambridge.
246. Tarnocai, C. 1974. Peatland forms and associated vegetation. In Proc. Canada Soil Survey Comm. Org. Soil Mapping Workshop, June 3–7, 1974, ed. J. H. Day, pp. 3–20 plus 20 fig. pp. Soil Research Institute, Central Experimental Farm, Ottawa.
247. Taylor, J. A., and R. T. Smith. 1980. Peat—a resource re-assessed. Nature 288:319–20.
248. Terasmae, J. 1977. Post-glacial history of Canadian muskeg. In Muskeg and the Northern Environment, ed. Radforth and Brawner, pp. 9–30. Univ. of Toronto Press.
249. Thompson, D. 1987. Battle of the bog. New Scientist, Jan. 8, pp. 41–44.
250. Thunmark, S. 1942. Über rezente Eisenocker und ihre Mikroorganismengemeinschaften. Akad. Avh. Bull. Geol. Inst. Uppsala 29:1–285.
251. Tinbergen, L. 1940. Observations sur l'evaporation de la végétation d'une

tourbière dans les Hautes-Fagnes de Belgique. Mém. Soc. Roy. Sci. Liège, 4th ser., 4:21–75.
252. Tolonen, K. 1971. On the regeneration of North European bogs. I. Klaukkalen Isosu in S. Finland. Acta Agralia Fenn. 123:143–66.
253. Tolonen, K. 1979. Peat as a renewable resource: Long-term accumulation rates in North European mires. *In* Proc. Intern. Symp. Hyytiälä, Finland, Sept. 17–21, 1979, pp. 282–96. International Peat Society, Helsinki.
254. Transeau, E. N. 1905–6. The bogs and bog flora of the Huron River valley. Bot. Gaz. 40:351–75, 418–48 (1905); 41:17–42 (1906).
255. Trappe, J. M. 1962. Fungus associates of ectotrophic mycorrhizae. Bot. Rev. 28:538–606.
256. Tuomikoski, R. 1946. Finnish *Sphagna* and their identification without a microscope [in Finnish]. Luonnon Ystava 50:113–17, 150–59.
257. Vitt, D. H., P. Achuff, and R. E. Andrus. 1975. The vegetation and chemical properties of patterned fens in the Swan Hills, north-central Alberta. Canad. Jour. Bot. 53:2776–95.
258. Vitt, D. H., and R. E. Andrus. 1977. The genus *Sphagnum* in Alberta. Canad. Jour. Bot. 55:331–57.
259. Vitt, D. H., and S. Bayley. 1984. The vegetation and water chemistry of four oligotrophic basin mires in northwestern Ontario. Canad. Jour. Bot. 62:1485–1500.
260. Vitt, D. H., H. Crum, and J. A. Snider. 1975. The vertical zonation of *Sphagnum* species in hummock-hollow complexes in northern Michigan. Mich. Bot. 14:190–200.
261. Vitt, D. H., and N. B. Slack. 1975. An analysis of *Sphagnum*-dominated kettlehole bogs in relation to environmental gradients. Canad. Jour. Bot. 53:332–59.
262. Vitt, D. H., and N. B. Slack. 1984. Niche diversification of *Sphagnum* relative to environmental factors in northern Minnesota peatlands. Canad. Jour. Bot. 62:1409–30.
263. Voice, S. P., M. R. Penskar, and J. P. Ludwig. 1984. The 2nd annual report on the ecological monitoring program of the Lake 16 *Sphagnum* peat project. Annual Rept. L16-83. Ecological Research Services, Inc., Boyne City, Mich.
264. Walker, D. 1961. Peat stratigraphy and bog regeneration. Proc. Linn. Soc. 172:29–33.
265. Walker, D., and P. M. Walker. 1961. Stratigraphic evidence of regeneration in some Irish bogs. Jour. Ecol. 49:169–85.
266. Warnstorf, C. 1890. Contributions to the knowledge of North American *Sphagna*. Bot. Gaz. 15: 127–40, 189–98, 217–27, 242–55.
267. Warnstorf, C. 1911. Sphagnales—Sphagnaceae (Sphagnologia Universalis). 51. Heft *In* Das Pflanzenreich, ed. A. Engler. Engelmann, Leipzig.
268. Watson, W. 1918. *Sphagna*, their habitats, adaptations, and associates. Ann. Bot. 32:533–51.
269. Watt, A. S. 1947. Pattern and process in the plant community. Jour. Ecol. 35:1–22.

270. Watts, W. A. 1983. Vegetational history of the eastern United States 25,000 to 10,000 years ago. *In* Late-Quaternary Environments of the United States, ed. S. C. Porter, pp. 294–310. Vol. 1, The Late Pleistocene. Univ. of Minnesota Press, Minneapolis.
271. Waughman, G. J., and D. J. Bellamy. 1980. Nitrogen fixation and the nitrogen balance in peatland ecosystems. Ecol. 61:1185–98.
272. Webb, T., III, E. J. Cushing, and H. E. Wright, Jr. 1983. Holocene changes in the vegetation of the Midwest. *In* Late-Quaternary Environments of the United States, ed. H. E. Wright, Jr., pp. 142–65. Vol. 2, The Holocene. Univ. of Minnesota Press, Minneapolis.
273. Weber, M. G., and K. van Cleve. 1981. Nitrogen dynamics in the forest floor of interior Alaska black spruce ecosystems. Canad. Jour. For. Research 11:743–51.
274. Weitzman, A. L. 1983. Summerby Swamp, an unusual plant community in Mackinac County, Michigan. Mich. Bot. 23:11–18.
275. Wells, E. D. 1981. Peatlands of eastern Newfoundland: Distribution, morphology, vegetation and nutrient status. Canad. Jour. Bot. 50:1978–97.
276. Wells, E. D., and F. C. Pollett. 1983. Peatlands. *In* Biogeography and Ecology of the Island of Newfoundland, ed. G. R. South, pp. 207–65. Junk, The Hague.
277. Wentz, W. A. 1976. The effect of simulated sewage effluents on the growth and productivity of peatland plants. Ph.D. diss., Univ. of Michigan.
278. West, R. G. 1977. Pleistocene Geology and Biology with Especial Reference to the British Isles. 2d ed. Longman, London and New York.
279. Witting, M. 1948. Forsatta Katjonbestamningar i Myroatten. Sv. Bot. Tidskr. 42:116–34.
280. Worley, I. A. 1980. Botanical and ecological aspects of coastal raised peatlands in Maine. Critical Areas Program, Planning Rept. 69. Maine State Planning Office, Augusta.
281. Worley, I. A. 1981. Maine Peatlands. Their abundance, ecology, and relevance to the Critical Areas Program of the State Planning Office. Critical Areas Program, Planning Rept. 73. Maine State Planning Office, Augusta.
282. Worley, I. A., and J. R. Sullivan. 1980. Classification scheme for the peatlands of Maine. Vt. Agr. Exp. Res. Rept. 8. Burlington, Vt.
283. Wright, H. E., Jr. 1976. The dynamic nature of Holocene vegetation. Quat. Res. (N.Y.) 6:581–96.
284. Zoltai, S. C. 1972. Palsas and peat plateaus in central Manitoba and Saskatchewan. Canad. Jour. For. Res. 2:291–302.
285. Zoltai, S. C., and F. C. Pollett. 1983. Wetlands in Canada: Their classification, distribution, and use. *In* Mires: Swamp, Bog, Fen and Moor, ed. A. J. P. Gore, pp. 245–68. Ecosystems of the World, vol. 4B. Elsevier, Amsterdam.
286. Zumberge, J. H., and J. E. Potzger. 1956. Late Wisconsin chronology of the Lake Michigan basin correlated with pollen studies. Bull. Geol. Soc. Amer. 67:271–88.

Index

Page numbers set in italics indicate illustrations.

Aapamire, 8, 9, *19*, *24*, *25*
  formation, *9*
Abies balsamea, 2, *18*, 65, *80*
Acer rubrum, 2, *17*, 68
Acetolysis, 166
Acid edge mat, *13*, 34, *52*, *53*, 64
  precipitation, 125, 155, 159, 160
Acidity, 143, 153–56
Aerobic peat, 116, 128
Airform patterns, 10, 11
Alder. *See* Alnus rugosa
Algae
  blooms, 142, 144
  blue-green, 119
  indicators, 144
Alkaline edge mat, *12*, 33, 59
Alnus rugosa, 3, *17*, 68, 119
American elm. *See* Ulmus americana
Amino acid uptake, 117
Ammonification, 117, 118
Ammonium uptake, 117
Anaerobic peat, 32, 116
Andromeda glaucophylla, *12*, 69, *79*, 87
Animal tracks, 134
Anion uptake, 181
Antibiosis, 194
Apron, 7, 20
Arceuthobium pusillum, 43, 62, 68, 130, *138*
Arethusa bulbosa, 67, *106*
Aronia prunifolia, 68, *75*, 86
Artifacts, 38, 167, 172
  Indian, 172
  Roman, 167

Viking, 167
Ash content, 180
Atmospheric
  dust, 32, 112
  pollutants, 159, 160
  temperature, 127, 128
Availability
  calcium, 126
  iron, 142
  nitrogen, 142
  nutrients, 113, 114, 124
  phosphorus, 142, 143

Bacteria, 115–24
Balsam fir. *See* Abies balsamea
Beach pool succession, 40, *54*, *55*
Beaver activity, *56*
Betula
  alleghaniensis, 2, *16*
  pumila, 3, *17*, 68
Black ash. *See* Fraxinus nigra
Black spruce. *See also* Picea mariana
  climax, 39–44
  islands, 10, 11, *25*
  layering, 34, *51*
  muskeg, 5, 11, *27*, *28*, 34, 35, 40, 43, *45*, *46*, 47, *51*, *52*, *62*, 63
Bladderworts. *See* Utricularia intermedia
Blanket bog, 8, *19*, 23, 31, 39, 123, 197
Bog, 1, 5, *12*, *13*, *19*, *21*, 23, 29, 31, *51–53*, 126
  blanket, 8, *19*, 23, 31, 39, 123, 197
  bodies, 172, 173

299

Bog (continued)
  burst, 151
  domed, 7, *19, 21,* 22
  hummock-hollow (open), 5, 11, 34, *51,* 62, 153
  iron, 123
  oak, 38
  pH, 5, 158
  raised, 3, 7
  succession, 32, 39–44
Boreal
  forest, *28, 29, 45, 46*
  peatlands, *28, 29, 45, 46*
Boundary layer, 167
Bronze Age, 37, 167, 172
Brown mosses, 54, 59
Buffering, 6, 32, 33, 60, 125, 155, 160, 192
Burned over bog, 43, 50, *56*
Butterworts. See Pinguicula vulgaris

Calamagrostis canadensis, 65, 101
Calcareous fens, *14,* 54, *55,* 153, 158
  soilwater limit, 6
Calciphiles, 36, 41, 54
Calcium, 6, 33, 42, 112, 126, 142, 156, 158, 160
  availability, 126
  bicarbonate buffering, 6, 32, 60, 125, 155, 160, 192
  cycling, 124
Calla palustris, 67, *108*
Calopogon tuberosus, 67, *106*
Canadian peat, 182
  Wetland Classification, 6
Carbon, 127
  dioxide, 127
  gradients, 128
  temperatures (climates), 127
Carbonate:bicarbonate equilibrium, 124
Carbon:nitrogen ratio, 117
Carex
  aquatilis, 66, *90*
  canescens, 66, *92*
  capillaris, 66, *94*
  cephalantha, 66, *96*
  chordorrhiza, 66, *95*
  comosa, 66, *90*
  diandra, 66, *92*
  disperma, 66, *93*
  exilis, 66, *96*

  flava, 66, *94*
  gynocrates, 66, *95*
  hystericina, 66, *72*
  interior, 66, *96*
  lacustris, 66, *90*
  lasiocarpa, 12, 34, 36, 47, 51, 58, 60, *93*
  leptalea, 66, *94*
  limosa, 66, *93*
  livida, 66, *92*
  oligosperma, 61, *91*
  pauciflora, 66, *91*
  paupercula, 6, *93*
  rostrata, 66, *91*
  sterilis, 66, *96*
  stricta, 66, *91*
  trisperma, 66, *93*
Carnivorous plants, 129, 130, *134–37*
Carr, 17, 36, 47
Cation exchange, 32, 121, 143, 153, 155–58, 161, 181
Cedar swamp, 11, *14,* 40–43, 59, 153
Cellulose, 115
Cephalanthus occidentalis, 2, 86
Chamaedaphne, 12, 32, 34, 35, 47, *51,* 62, 69, 76, 87, 114
Chelation, 121, 123, 142, 145, 155
Cladium mariscoides, 66, *99*
Climax, 39–44
Cold air drainage, 145, 146
Collapse scar, 10, *26*
Color
  peatland water, 144
  Sphagnum, 149
Complexing. See chelation
Conservation, 196–99
Corallorhiza, 133
Cornus stolonifera, 2, 69, 86
Cypripedium
  acaule, 67, *107*
  calceolus, 67, *107*

Decodon verticillatus, 2, *84*
Deer trails, 112, 134
Denitrification, 118
Dionaea, 129, *137*
Dissociation of organic acids, 143, 155
Diversity, 31, 33, 34, 61
Domed bogs, 7, *19, 21,* 22
  concentric, 7, *19, 21*
  eccentric, 7, *19,* 22

Drosera, 129, *135*, *136*
  intermedia, 68, *135*
  linearis, 68, *135*
  rotundifolia, 68, *135*
Drought, 34, 50, 51, 144, 150, 154
Drunken forest, 10
Dulichium arundinaceum, 66, 72, 99
Dwarf birch. *See* Betula pumila
Dwarf mistletoe. *See* Arceuthobium pusillum
Dystrophic lakes, 144, 145

Ectomycorrhizae, 131, *139*
Electricity generation, 183
Eleocharis elliptica, 66, *100*
Empetrum nigrum, 59, 68, *89*
Endomycorrhizae, 132, *139*
Ericaceae, 120
  adaptations, 132, 145
Eriocaulon septangulare, 67, 72
Eriophorum
  angustifolium, 66, *98*
  spissum, 67, *98*
  virginicum, 67, *71*, *98*
  viridi-carinatum, 67, *71*, *98*
Ethanol production, 128
Euminerotrophic mire, 4
Eutrophic, 4, 111, 165
  lakes, 144
  mire, 4
Evapotranspiration, 39, 145, 148–50
Evergreen leaves, 114, 120, 129
Extreme poor fen, 5

False bottom, 12, 33, 34, 36, *50*, *51*, *54*, 60, 144, 150
  exposure, *50*, *51*, *54*
Fen, 1, 5, *12*, *14*, 17, 29, 32, 37, *50*, *54*, 55, 111, 126
  calcareous, *14*, *54*, *55*, 153, 158
  intermediate, 5, 11, *12*, 33, 44, *50*, 60, 153
  poor, 4, 5, 11, 17, 31, 34, 35, 61, 153
  rich, 4, 5, 11, *14*, 32, 36, 40, 41, *54*, 55, 57, 153
  succession, 42, 44, 59
  treed, 2
Fertilizing peat, 189, 192
Fiber content, 180
Fibric peat, 150, 158, 177, 179–82, 186

Fire, 34, 43, 44, 50, *56*, 191
Flachmoor, 4
Flarks, 8, 9, *24*, *25*
Flat bog, 4
Floating mats, 34, 120, 152
Flooding, 43, *56*
Fog, 145, *146*
Fraxinus nigra, 2, *16*
Frost, 146, 148, 190
  heaving, 148, 190
  polygons, 11, *28*
Fulvic acid, 155
Fungi, 133

Galacturonic acid, 155–58
Gaultheria hispidula, 69, *74*, *89*
Gaylussacia baccata, 69, *88*
Glacial history, 163–73
  Lake Agassiz, 38
  landforms, 30, 163, 164
Glyceria
  canadensis, 65, *102*
  striata, 65, *71*, *102*
Glycolysis, 128
Golden moss, 225
Greenhouse effect, 127
Grenzhorizont, 167
Grounded mat, 34, 35
Groundwater limit, 6
Gyttja, 33, 142

Habenaria
  blephariglottis, 67, *105*
  clavellata, 67, *105*
Hardpan, 8, 123
Heat
  absorption, 147
  radiation, 145
  traps, 147
Hemicellulose, 115
Hemic peat, 150, 158, 177, 179, 180, 186
High moor, 4
Hiller peat sampler, 165, *174*
Hinge line, 34
Hochmoor, 4
Huckleberries, 57, 88
Hudson Bay Lowlands, *28*, 45
Humic
  acids, 142, 155

Humic (*continued*)
  colloids, 142
  compounds, 144
  substances, 155–81
Humification values, 178, 179
Humin, 155
Hummock
  formation, 34
  pH gradient, 159
  water-holding, 149
  zonation, 149
Hummock-hollow bog, 5, 11, 34, 62, 153

Ice push ridge, 34, *47*
Ilex verticillata, 2, 68, *85*
Indian artifacts, 172
Indicators, 32, 33, 57–70
  algae, 144
  black spruce muskeg, 63
  calcium, 31, 36, 40, 57–59, 153
  hummock-hollow bog, 62
  intermediate fen, 60
  lagg, 63
  poor fen, 61
  rich fen, 58
Intermediate fen, 5, 11, *12*, 33, 44, *50*, 60, 153
Intermediate fen–black spruce successions, 44
Iris versicolor, 67, *109*
Iron, 142
  availability, 142
  chelation, 142
  cycling, 123, 124, 143
  ferric, 123, 143
  ferrous, 122, 123, 143
  pans, 8, 123
Iron Age, 37, 167, 172
Itchy muck, 193

Jack pine. *See* Pinus banksiana
Juncus balticus, 67, *100*

Kalmia polifolia, 69, 76, *87*
Karr, 4
von Keppler humification values, 179
Kettlehole lake, *12*, 30

Lagg, 7, 17, *20, 21,* 34, 35, *52,* 59, 63
  indicators, 63

Lake environment, 141–43
Lake-fill bog, *19, 20,* 29, 32, 34–36, *47–49, 53*
  mapped, *53*
Larix laricina, 2, *18,* 65, *79*
Lawn dressing, 178, 182
Ledum groenlandicum, 35, 69, 76, *87*
Lignin, 115, 177
Lindow man, 172
Liparis loeselii, 67, 105
Lonicera oblongifolia, 70, *85*
Low moor, 4

Macrofossils, 164
Macronutrients, 112
Magnesium, 33, 127
Malate production, 128
Marl, *14,* 33, *54,* 124, 125
Marsh, 1, *13*
  grass, 91
  hay, 91
Menyanthes trifoliata, 69, *109*
Mesotrophic mire, 4
Methane production, 116
Michigan peat, 182
Micronutrients, 112
Mineral-rich lake, 33
Minerals. *See* Availability; Nutrients
Mineral uptake, 131
Minerotrophic lake mat, 36
Moat. *See* lagg
Monotropa, 133, 140
Moss (bog), 4, 5
Muck, 148, 182, 190–93
  fires, 191
Muhlenbergia glomerata, 65, *101*
Muskeg, 5, 11, *27, 28,* 34, 35, 40, 43, *45, 46, 47,* 62, 63
Mycorrhizae, 130–33, *139,* 140
Myrica gale, 68, *75, 81,* 119, 120

Natural gas, 183
Nemopanthus mucronatus, 68, *74, 85*
Niedermoor, 4
Nitrate uptake, 117
Nitrification, 118
Nitrogen, 112, 115
  availability, 115
  cycle, 115–20, 142
  fixation, 75, 112, 118–20

Nitrogenous decay, 116, 121
Nutrient
  availability, 113, 114
  cycling, 111, 114–20, 129, 143, 160, 195, 196
  translocations, 120, 126, 129
  uptake, 113, 120, 121, 131, 181
Nutrients, 112–21, 129, 134, 143, 160, 181, 191
Nymphaea odorata, 71

Oceanic climates, 197
Oligotrophic, 4, 111, 165
  lakes, 144
  mires, 4
Ombrotrophic mire, 4, 5, 17
Onoclea sensibilis, 65, 78
Open bog, 5, 11, 31, 51, 62, 153
Orchidaceae, 132
Organic acid dissociation, 143, 155
Osmunda
  cinnamomea, 65, 77
  regalis, 65, 77
Oxygen, 32, 33, 116, 128, 143
  deficiency, 32, 116, 128
  gradient, 128

Palsa, 10, 26, 28
  mire, 10, 28
Paludification, 29, 30, 37–39, 48, 49, 56, 64, 164
Parasitism, 130, 138
Parnassia glauca, 68, 104
Patterned peatland. See aapamire
Peat
  accumulation rate, 181
  air capacity, 181
  aquatic, 177
  ash content, 180, 192
  block cut, 182
  briquettes, 182
  bulk density, 180
  Canadian, 182
  classification, 177–78
  compressibility, 181
  coring, 165
  defined, 177
  deposits, 37
  depth, 184
  farming, 189–93

fibric, 150, 158, 177, 179–81
fuel, 178, 179
hand cut, 182
heat capacity, 147
hemic, 150, 158, 177, 179, 180
herbaceous, 177
horticultural, 178, 179, 181
humification, 178, 179
impermeability, 32, 151
insulation, 147
Michigan, 182
milled, 182, 186
mining, 182–88
pH, 158, 159
plateau, 10, 27, 28
porosity, 18
properties, 177–81, 190–92
reed-sedge, 182
reserves, 181–83
ridge, 10
sampler, 165, 174
sapric, 150, 158, 177, 179–81
sedge, 177, 182
Sphagnum, 177, 179, 181
substitutes, 192
uses, 177–84, 189–96
water-holding capacity, 150, 180, 181
wood, 177
Peatland
  agriculture, 189–93
  classification, 3–9
  climax, 39–44
  conservation, 196–99
  defined, 1
  drainage, 147, 148, 151
  patterned, 8, 9
  plants, 65–70
  primary, 3
  reclamation, 198
  regeneration, 196–99
  secondary, 3
  temperature, 145, 147
  tertiary, 3
Peat soil (muck)
  aeration, 191, 192
  burning, 191
  draining, 147, 148, 151, 190
  drying, 191, 192
  fertilizing, 192
  frost heaving, 148, 190

Peat soil (continued)
  nutrients, 190, 192
  subsidence, 190, 191
  temperature, 147, 148, 181, 190, 191
  water storage and release, 151
Permafrost, 10, 11, 29, 39
pH, 5, 119, 123, 158–60
  bog, 5, 158
  fen, 5, 158
  gradient, 153, 157–59
  hummocks, 158, *159*
  nutrient availability, 113–15, 126
  peat, 158, *159*
  rainwater, 125, 159
  seasonal changes, 157
Phosphorus
  availability, 120, 142, 143
  cycle, 143
Phragmites australis, 65, 73, *101*
Picea
  glauca, 2, 65, *79*
  mariana, 2, *18, 28,* 43, *55,* 65, *79*
Pingo, 11
Pinguicula vulgaris, 69, 129, *135,* 137
Pin oak. *See* Quercus palustris
Pinus
  banksiana, 41, *55,* 65, *79*
  strobus, 41, *55*
Pitcher plant. *See* Sarracenia purpurea
Plateau raised bog, 7, *19–21*
Podzolization, 39
Pogonia ophioglossoides, 67, *106*
Pollen analysis, 37, 164–67, 170–72
  diagrams, 167, 172, *175, 176*
Pond porridge, *14,* 125
Poor fen, 4, 5, 11, 17, 31, 34, 35, 61, 153
  pH, 5
Postglacial
  climates, 37, 164, 166–72
  vegetation, 164, 166, 167, 169–72
von Post humification scale, 178, 179
Potassium, 120
  leaching, 121
Potentilla
  fruticosa, 68, *84*
  palustris, 68, *84*
Potting soil, 178, 182
Precipitation:evaporation balance, 8, 29, 146
Productivity, 111, 143, 144
Pyrola asarifolia, 69, *104*

Quercus palustris, 2, *15*

Rain, 31, 112, 113, 157, 197
  pH, 125
Raised bog, 3, 7
Rand, 7
Reclamation of peatlands, 198
Recurrence surfaces, 154, 167
Red maple. *See* Acer rubrum
Reed-sedge peat, 182, 183
Regeneration cycle, 35, *151,* 154
Regeneration of peatlands, 196–99
Rhamnus alnifolia, 68, *83*
Rheophilous mire, 4, 5
Rhynchospora alba, 67, *97*
Rich fen, 4, 5, 11, *14,* 32, 36, 40, 41, *54,* 55, 57, 153
  pH, 5
Rich fen–cedar swamp succession, 41, 42, 44, *55,* 59
Riekarr, 4
Roman artifacts, 38, 167
Root functions, 147, 148
Rubus pubescens, 68, *74*

Salix
  candida, 67, *81*
  pedicellaris, 68, *81*
  petiolaris, 68, *81*
  pyrifolia, 68, *82*
  serissima, 68, *82*
Sapric peat, 150, 158, 177, 179–81
Sarracenia purpurea, 68, 129, 130, *134, 137*
Scheuchzeria palustris, 65, *72, 103*
Scirpus
  acutus, 67, *97*
  cespitosus, 23, 67, *98*
  cyperinus, 67, *97*
  hudsonianus, 67, *72, 98*
  validus, 67, *97*
Seasonal overturns, 142, 143
Sedge fen, 5, 12, *50*
Sedge mat, 34, 47, 60, 144
  advance, 34, *50, 51,* 144
  pioneer, 34, 153
Sedge meadow, 1
Sedge peat, 182, 183, 192
Sediments, 33, 142, 144, 164, 165, 167
Seepage force, 9
Sewage treatment, 195, 196

Silviculture, 189, 197
Slope bog, 8, *21*
Smilacina trifolia, 67, *108*
Snowmelt, 112
Snow pack, 147, *149*
Soligenous, 4
  fen, 7
Sphagnol, 115
Sphagnum, miscellaneous features
  acidification, 153, 154, 157
  antibiosis, 194
  cation exchange, 153, 157
  color, 149
  ecological role, 149, 152–54, 157, 161
  heat absorption, 149
  identification, 201–41
  indicators, 61
  key to sections, 221
  lawn, 5, 35, 36, 153
  life history, 201
  peat, 180, 192
  pioneer species, 35, 153
  role in succession, 30
  staining, 217
  temperature, 49, 161
  uses, 193–96
  water relations, 152, 153, 161
Sphagnum sections and species
  sect. Acutifolia, 222, 236
    capillifolium, 237, *245, 267*
      var. tenellum, *209,* 238
      var. tenerum, 238
    fimbriatum, 209, 241, *246, 275*
    flavicomans, *271*
    fuscum, 55, 239, *246, 269*
    girgensohnii, *209,* 241, *246, 274*
    molle, *212*
    nemoreum, 237
    quinquefarium, 240, *272*
    rubellum, 238
    russowii, 208, *210,* 240, *246, 273*
    subfulvum, 239, *271*
    subnitens, 239, *270, 271*
    subtile, 238
    tenerum, 238
    warnstorfii, 55, *212,* 239, *246, 268*
  sect. Cuspidata, 222, 228
    angustifolium, 230
    annulatum var. prosum, 232
    cuspidatum, *212,* 231, *245, 261*
      var. serrulatum, *214*
    fallax, 229
    flexuosum, 229
    jensenii, 232, *263*
    lindbergii, *209*
    majus, 212, 232, *244, 262*
    obtusum, *211, 212,* 230, *259*
    pulchrum, *214,* 230, *244, 258*
    recurvum, 212
      var. brevifolium, 229, *243, 257*
      var. tenue, 230
    riparium, *209,* 231, *244, 260*
    splendens, *212,* 230
    tenellum, *208*
    torreyanum, 232, *245*
  sect. Hemitheca, 221, 233
    pylaesii, 233, *264*
  sect. Insulosa, 222, 226
    aongstroemii, 226, *252*
  sect. Isocladus, 221, 227
    macrophyllum, 227, *243, 255, 256*
      var. burinense, 228, *256*
      var. floridanum, 227, *256*
  sect. Polyclada, 222, 235
    wulfianum, 235, *245, 266*
  sect. Rigida 221, 225
    compactum, *215,* 225, *242, 251*
  sect. Sphagnum, 221, 223
    centrale, *208, 214, 223, 242, 248*
    henryense, *210, 211, 213,* 224
    imbricatum, *211, 213,* 224
    magellanicum, 133, *210, 213,* 223, *242, 247*
    palustre, *211,* 224, *249*
    papillosum, *210, 213, 214,* 224, *242, 250*
    portoricense, *208, 210, 212, 215*
  sect. Squarrosa, 222, 233
    squarrosum, 227, *242, 253*
    teres, 33, 153, *209,* 227, *242, 254*
  sect. Subsecunda, 221, 233
    contortum, 235
    lescurii, 234
    platyphyllum, 235
    subsecundum, *213,* 234, *245, 265*
      var. contortum, 235
      var. inundatum, 234
      var. platyphyllum, 235
      var. rufescens, 234, *245*
Sphagnum structure, 202–5, 216
  antheridia, 204, *215*
  commissural pores, 204, *213*

Sphagnum structure (continued)
  cortex, 203, 208
  divided cells, 203, 210
  fibrils, 204, 208, 210–12
  green cells, 204, 211, 214
  leaf border, 204, 214, 215
  membrane gaps, 203, 204, 211, 213
  membrane pleats, 203, 210
  ornamented walls, 204, 212–14
  pores, 204, 208, 210–13
  protonema, 202, 206, 207
  pseudopodium, 205, 215
  resorption, 203, 204, 209–11, 214, 215
  resorption furrow, 204, 214, 215
  retort cells, 203, 208
  spore, 202, 206
  stem leaves, 203, 209
Spiraea alba, 68, 73, 84
Sporotrichosis, 193
Stratigraphy. See Pollen analysis
String, 8, 9, 24, 25
String bog. See Aapamire
Succession, 39–44, 59, 64, 148, 164, 166, 168–72
Sulfur cycling, 121, 122, 143, 155
Sundews. See Drosera
Swamp, 2, 11, 14, 17, 40–43, 59, 153

Tamarack. See Larix laricina
Tannin, 144, 145
Teardrop island, 9, 10, 25
Temperature
  carbon dioxide, 127, 129
  diurnal fluctuations, 145–47
  drainage, 147
  peat, 181
  root functions, 147, 148
Thelypteris palustris, 65, 78
Thuja occidentalis, 2, 18, 40–43, 59, 65, 80, 153
Tilletia, 206, 216
Tofieldia glutinosa, 67, 104
Tollund man, 172
Topography of peatlands, 19
Toxicodendron vernix, 68, 83
Transitional
  poor fen, 5
  rich fen, 5
Transition
  mire, 4
  moor, 4

Triglochin
  maritimum, 65, 103
  palustre, 65, 103
Typha latifolia, 52, 59, 65

Uebergangsmoor, 4
Ulmus americana, 2, 15
Uses of peat, peatlands, and peat mosses, 177–96
Utricularia intermedia, 35, 69, 129, 136

Vaccinium
  angustifolium, 69, 88
  corymbosum, 69, 88
  macrocarpon, 69, 89
  myrtilloides, 69, 74, 88
  oxycoccus, 69, 89
Vacuum harvesting, 186, 187, 188
Venus's fly trap. See Dionaea
Viburnum cassinoides, 70, 75, 83
Viking artifacts, 167

Waste water treatment, 195, 196
Water
  absorption, 148
  capillary, 149, 150
  content, 180, 181
  holding, 148
  level, 33, 38, 144, 152
  loss, 148
  stress, 145, 148
  table, 7, 150, 151
  track, 112
  upwelling, 152
Waterlogging, 128
Weakly minerotrophic mire, 4
Whisky, 194, 195
White cedar. See Thuja occidentalis
White pine. See Pinus strobus
White spruce. See Picea glauca
Will-o'-the-wisp, 116
Witches' brooms, 130

Xeromorphy, 145
Xyris
  diffusa, 99
  montana, 67, 99

Yellow birch. See Betula alleghaniensis

Zwischenmoor, 4